Multicomponent Pharmaceutical Solids

Multicomponent Pharmaceutical Solids

Editors

**Duane Choquesillo-Lazarte
Alicia Domínguez Martín**

MDPI • Basel • Beijing • Wuhan • Barcelona • Belgrade • Manchester • Tokyo • Cluj • Tianjin

Editors
Duane Choquesillo-Lazarte
Laboratorio de Estudios
Cristalográficos, IACT, CSIC
Granada
Spain

Alicia Domínguez Martín
University of Granada
Granada
Spain

Editorial Office
MDPI
St. Alban-Anlage 66
4052 Basel, Switzerland

This is a reprint of articles from the Special Issue published online in the open access journal *Crystals* (ISSN 2073-4352) (available at: https://www.mdpi.com/journal/crystals/special_issues/crystals_pharmaceuticalsolids).

For citation purposes, cite each article independently as indicated on the article page online and as indicated below:

LastName, A.A.; LastName, B.B.; LastName, C.C. Article Title. *Journal Name* **Year**, *Volume Number*, Page Range.

ISBN 978-3-0365-7216-1 (Hbk)
ISBN 978-3-0365-7217-8 (PDF)

Cover image courtesy of Duane Choquesillo-Lazarte

© 2023 by the authors. Articles in this book are Open Access and distributed under the Creative Commons Attribution (CC BY) license, which allows users to download, copy and build upon published articles, as long as the author and publisher are properly credited, which ensures maximum dissemination and a wider impact of our publications.

The book as a whole is distributed by MDPI under the terms and conditions of the Creative Commons license CC BY-NC-ND.

Contents

About the Editors vii

Preface to "Multicomponent Pharmaceutical Solids" ix

Duane Choquesillo-Lazarte and Alicia Domínguez-Martín
Multicomponent Pharmaceutical Solids
Reprinted from: *Crystals* 2023, 13, 570, doi:10.3390/cryst13040570 1

Yanan Wang, Yong Wang, Jin Cheng, Haibiao Chen, Jia Xu, Ziying Liu, Qin Shi, et al.
Recent Advances in the Application of Characterization Techniques for Studying Physical Stability of Amorphous Pharmaceutical Solids
Reprinted from: *Crystals* 2021, 11, 1440, doi:10.3390/cryst11121440 5

Cristina Puigjaner, Anna Portell, Arturo Blasco, Mercè Font-Bardia and Oriol Vallcorba
Entrapped Transient Chloroform Solvates of Bilastine
Reprinted from: *Crystals* 2021, 11, 342, doi:10.3390/cryst11040342 23

Qi Zhou, Zhongchuan Tan, Desen Yang, Jiyuan Tu, Yezi Wang, Ying Zhang, Yanju Liu, et al.
Improving the Solubility of Aripiprazole by Multicomponent Crystallization
Reprinted from: *Crystals* 2021, 11, 343, doi:10.3390/cryst11040343 45

Francisco Javier Acebedo-Martínez, Carolina Alarcón-Payer, Lucía Rodríguez-Domingo, Alicia Domínguez-Martín, Jaime Gómez-Morales and Duane Choquesillo-Lazarte
Furosemide/Non-Steroidal Anti-Inflammatory Drug–Drug Pharmaceutical Solids: Novel Opportunities in Drug Formulation
Reprinted from: *Crystals* 2021, 11, 1339, doi:10.3390/cryst11111339 61

Basanta Saikia, Andreas Seidel-Morgenstern and Heike Lorenz
Multicomponent Materials to Improve Solubility: Eutectics of Drug Aminoglutethimide
Reprinted from: *Crystals* 2022, 12, 40, doi:10.3390/cryst12010040 75

Mónica Benito, Miquel Barceló-Oliver, Antonio Frontera and Elies Molins
Oxalic Acid, a Versatile Coformer for Multicomponent Forms with 9-Ethyladenine
Reprinted from: *Crystals* 2022, 12, 89, doi:10.3390/cryst12010089 87

Alfonso Castiñeiras, Antonio Frontera, Isabel García-Santos, Josefa M. González-Pérez, Juan Niclós-Gutiérrez and Rocío Torres-Iglesias
Multicomponent Solids of DL-2-Hydroxy-2-phenylacetic Acid and Pyridinecarboxamides
Reprinted from: *Crystals* 2022, 12, 142, doi:10.3390/cryst12020142 103

Laura Baraldi, Luca Fornasari, Irene Bassanetti, Francesco Amadei, Alessia Bacchi and Luciano Marchiò
Salification Controls the In-Vitro Release of Theophylline
Reprinted from: *Crystals* 2022, 12, 201, doi:10.3390/cryst12020201 125

Cristóbal Verdugo-Escamilla, Carolina Alarcón-Payer, Francisco Javier Acebedo-Martínez, Raquel Fernández-Penas, Alicia Domínguez-Martín and Duane Choquesillo-Lazarte
Lidocaine Pharmaceutical Multicomponent Forms: A Story about the Role of Chloride Ions on Their Stability
Reprinted from: *Crystals* 2022, 12, 798, doi:10.3390/cryst12060798 135

Francisco Javier Acebedo-Martínez, Carolina Alarcón-Payer, Helena María Barrales-Ruiz, Juan Niclós-Gutiérrez, Alicia Domínguez-Martín and Duane Choquesillo-Lazarte
Towards the Development of Novel Diclofenac Multicomponent Pharmaceutical Solids
Reprinted from: *Crystals* **2022**, *12*, 1038, doi:10.3390/cryst12081038 . . . . **151**

About the Editors

Duane Choquesillo-Lazarte

Duane Choquesillo-Lazarte received his PhD in Pharmacy in 2006 from the University of Granada. Being a pharmacist, he was interested in crystallography and soon recognized the extraordinary potential that the structure–property paradigm has in the rational development of high-quality research and, therefore, the relevant role that chemical crystallography has in designing new molecules based on structural knowledge. He has been a senior research scientist at the Laboratorio de Estudios Cristalográficos at the Andalusian Institute of Earth Sciences, CSIC in Granada, Spain since 2017. His research focuses on the design of multicomponent molecular materials (i.e. pharmaceutical cocrystals) in order to overcome biopharmaceutical limitations and drug manufacturing constraints. He also has a particular interest in the development of experimental/virtual cocrystal screening methods and crystallization methods using supramolecular gels. This research area stems from a fundamental interest in understanding intermolecular interactions in molecular systems.

Alicia Domínguez Martín

Alicia Domínguez Martín (Dr.) received her PhD in Pharmacy in 2013 from the University of Granada. From 2013 to 2017 she was a postdoctoral researcher at the Department of Chemistry at the University of Zurich. During this period, she delved into the study of molecular recognition patterns between metal ions and nucleic acids, using high-resolution techniques to unravel their intimate structure She has been an Associate Professor at the University of Granada since 2017. Her research interests currently involve the development of two different projects: (1) Metal–DNA nanobiomaterials; and (2) Multicomponent pharmaceutical solids. Regarding the latter, her main research activities are focused on enhancing the oral bioavailability of active pharmaceutical ingredients by designing, synthesizing, and characterizing novel drug–drug pharmaceutical solids that improve their original physicochemical properties keeping APIs' efficacy.

Preface to "Multicomponent Pharmaceutical Solids"

Multicomponent pharmaceutical materials are solids in which at least one component is an active pharmaceutical ingredient (API). This kind of pharmaceutical solid has certainly revolutionized the pharmaceutical industry, proving to be an interesting alternative to the laborious and expensive traditional pipeline drug development. The application of crystal engineering techniques into the design of pharmaceutical salts, co-crystals, and other multicomponent materials has succeeded in modulating the physicochemical, mechanical, and pharmacokinetic properties of drugs, thereby working towards the enhancement of their clinical performance. In addition, environmentally friendly synthetic approaches have also been achieved, in agreement with the new green chemistry principles.

Despite its promising future, the study of pharmaceutical solids still presents significant challenges. For instance, unraveling structure–activity relationships or the preference for supramolecular synthons, and choosing an appropriate co-former and/or solvent selection are still issues that need to be thoroughly studied in order to fully understand, and consequently predict, the formation of pharmaceutical solids with tailored properties.

We are delighted to edit this Special Issue called "Multicomponent Pharmaceutical Solids". New multicomponent pharmaceutical materials have come to stay, and we will soon see relevant applications on multidrug resistance or co-drug synergy thanks to their rational design. Until then, this Special Issue is devoted to compiling all current aspects related to the field of multicomponent pharmaceutical solids. We hope that this volume is of interest to all *Crystals* readers.

Duane Choquesillo-Lazarte and Alicia Domínguez Martín
Editors

Editorial

Multicomponent Pharmaceutical Solids

Duane Choquesillo-Lazarte [1,*] and Alicia Domínguez-Martín [2,*]

1. Laboratorio de Estudios Cristalográficos, IACT, CSIC-Universidad de Granada, Avda. de las Palmeras 4, 18100 Armilla, Spain
2. Department of Inorganic Chemistry, Faculty of Pharmacy, University of Granada, 18071 Granada, Spain
* Correspondence: duane.choquesillo@csic.es (D.C.-L.); adominguez@ugr.es (A.D-M.)

Multicomponent pharmaceutical solids is a hot topic that brings together the knowledge of crystal engineering and the need to achieve novel and effective drugs at lower costs for the pharmaceutical industry. Both the U.S. Food and Drug Administration and the European Medicines Agency encourage the use of pharmaceutical cocrystals for drug development, which is tightly regulated with appropriate guidelines. Indeed, in the past years, several cocrystal formulations have been approved and commercialized, for instance, Depakote® (2002), Suglat® (2014), Entresto® (2015), Steglatro® (2018), and Seglentis® (2021).

This issue collects 10 contributions. The first consists of a review article dealing with amorphous pharmaceutical solids [1]. Most of the multicomponent pharmaceutical solids reported are in crystalline form since they are thermodynamically more stable. Nonetheless, amorphous materials usually present higher solubility and a faster dissolution rate, hence being worth exploring due to improved oral bioavailability. In this context, Q. Shi, C. Zhang, et al. review the latest progress regarding the characterization of the physical stability of amorphous pharmaceutical solids.

The additional nine contributions are original research papers. First, C. Puigjaner et al. [2] compared the thermodynamic stability of three anhydrous polymorphs and thoroughly characterized three transient chloroform solvates of Bilastine, a second-generation antihistamine drug. Interestingly, the detailed structural study provides new insights into hydrate/solvate formation, paying special attention to the ability to build hydrogen bonding networks.

The physicochemical properties of pharmaceutical solids, such as solubility, thermal stability, and hygroscopicity, are highly dependent on their intimate solid-state structure. Hence, it is possible to modulate the pharmacokinetic profile of active pharmaceutical ingredients (APIs) by tailoring their crystal architecture. This is actually desirable for more than 80% of the drugs already on the market, which belong either to classes II or IV within the Biopharmaceutical Classification System (BSC), i.e., those exhibiting poor aqueous solubility. Indeed, the vast majority of the reported multicomponent pharmaceutical solids are devoted to enhancing the solubility of already-known APIs.

This is the case of the results reported by Y. Liu, G. Gan, et al., regarding the third-generation antipsychotic Aripiprazole [3]. The authors show new potential multicomponent formulations for this drug, highlighting the improved solubility of the corresponding salt with the coformer adipic acid. Single crystal X-ray diffraction and theoretical methods are employed to explore molecular interactions, while additional physicochemical properties are evaluated.

Fixed-dose formulations refer to multicomponent pharmaceutical solids in which all components are APIs. D. Choquesillo-Lazarte et al. [4] report the synthesis and characterization of furosemide/non-steroidal anti-inflammatory drug–drug pharmaceutical cocrystals, a combination of APIs that is quite frequent in clinics. Remarkably, the mechanochemical synthetic approach allowed for the avoidance of the formation of hydrates/solvates. Although solubility enhancement is not achieved in the novel cocrystals, improved thermal and thermodynamic stability profiles are given, opening the door to further research.

Citation: Choquesillo-Lazarte, D.; Domínguez-Martín, A. Multicomponent Pharmaceutical Solids. *Crystals* **2023**, *13*, 570. https://doi.org/10.3390/cryst13040570

Received: 21 March 2023
Revised: 22 March 2023
Accepted: 22 March 2023
Published: 27 March 2023

Copyright: © 2023 by the authors. Licensee MDPI, Basel, Switzerland. This article is an open access article distributed under the terms and conditions of the Creative Commons Attribution (CC BY) license (https://creativecommons.org/licenses/by/4.0/).

Eutectic mixtures are frequently used in the pharmaceutical industry regarding research on drug delivery systems. B. Saikia et al. [5] explore the use of such eutectic mixtures to improve the solubility of the non-steroidal aromatase inhibitor aminoglutethimide, using three different non-isomorphous coformers: caffeine, nicotinamide, and ethenzamide. Solvent-assisted mechanochemical synthesis was also employed to obtain the novel binary eutectic mixtures, which were thoroughly characterized and proved to successfully enhance the solubility of the parent API.

In M. Benito et al.'s paper [6], efforts were made to rationalize the nature and strength of those supramolecular interactions present in six new multicomponent systems involving the modified nucleobase 9-ethyladenine and the coformer oxalic acid. Herein, salt screening was carried out by mechanochemical and solvent crystallization techniques, yielding two molar ratios and different anhydrous/solvated forms. Comprehensive computational analysis helps to fully understand the solid-state landscape, unveiling the relative importance of non-canonical hydrogen bonds.

Molecules included in the Generally Regarded As Safe (GRAS) list are used along with the drug precursor DL-mandelic acid to design novel multicomponent pharmaceutical solids in A. Castiñeiras et al.'s paper [7]. The authors focus their attention on the identification of recurrent supramolecular patterns thanks to detailed crystal structure and computational analyses. Solubility studies were also performed, which revealed an enhancement of the dissolution profile.

Drugs with a narrow therapeutic index, such as theophylline, are better administered with sustained-release formulations, which reduce the likelihood of adverse reactions. L. Marchiò et al. [8] address the optimization of theophylline pharmacokinetics by using the salification strategy. The authors successfully decreased the dissolution rate of the API, which might allow better control of drug release into the bloodstream, therefore reducing adverse effects without losing drug efficacy.

The salification approach can also be observed in the formulation of the anesthetic drug Lidocaine, which is found as the corresponding hydrochloride derivative in order to improve the solubility of the API. In C. Verdugo-Escamilla and A. Domínguez-Martín et al.'s paper [9], ionic cocrystals of lidocaine base and lidocaine hydrochloride are reported along with polyphenols as coformers. Interestingly, only those multicomponent solids involving lidocaine hydrochloride achieved improved stability thanks to a steric protection effect, in which chloride ions play a key role in promoting efficient packing, thus protecting the coformer from oxidation.

The last contribution of this Special Issue is devoted to multicomponent formulations of the non-steroidal anti-inflammatory drug Diclofenac, using nucleobases as coformers due to their versatility to build multiple H-bonding patterns [10]. Careful structure-property analysis reveals that the large surface exposure of two of the novel cocrystals favors their dissociation in aqueous media. The third species undergoes dissociation at a much lower rate, therefore allowing its solubility to be characterized with positive results.

In summary, the articles presented in this Special Issue represent some of the most recent research approaches to the study of multicomponent pharmaceutical solids. The vast majority of results are focused on crystalline solids. The combination of X-ray crystallography and computational methods provides a complementary and comprehensive view of the nature and strength of supramolecular synthons, which drive the crystal structure architecture and therefore the physicochemical properties of the novel multicomponent materials. Solubility and dissolution rate, as well as thermal and thermodynamic stability, are the most common properties assessed, in agreement with the limitations currently observed in the drugs already present in the pharmaceutical market. The latest advances in multicomponent pharmaceutical solids are devoted not only to obtaining API + GRAS-like coformer formulations but also API + API formulations, so-called fixed-dose formulations. Likewise, multicomponent pharmaceutical solids' research goes beyond cocrystals, and molecular salts, ionic cocrystals, cocrystal polymorphs, and even amorphous phases are currently being evaluated. There are still significant challenges to overcome. In this context, further research on the development of structure-property studies is of paramount importance to the design of tailored properties and the future of multicomponent drugs.

Funding: Grant number PGC2018-102047-B-I00 (MCIU/AEI/FEDER, UE) and B-FQM-478-UGR20 (FEDER-Universidad de Granada, Spain).

Acknowledgments: The guest editors thank all the authors contributing to this Special Issue and the editorial staff of *Crystals* for their support.

Conflicts of Interest: The authors declare no conflict of interest.

References

1. Wang, Y.; Wang, Y.; Cheng, J.; Chen, H.; Xu, J.; Liu, Z.; Shi, Q.; Zhang, C. Recent Advances in the Application of Characterization Techniques for Studying Physical Stability of Amorphous Pharmaceutical Solids. *Crystals* **2021**, *11*, 1440. [CrossRef]
2. Puigjaner, C.; Portell, A.; Blasco, A.; Font-Bardia, M.; Vallcorba, O. Entrapped Transient Chloroform Solvates of Bilastine. *Crystals* **2021**, *11*, 342. [CrossRef]
3. Zhou, Q.; Tan, Z.; Yang, D.; Tu, J.; Wang, Y.; Zhang, Y.; Liu, Y.; Gan, G. Improving the Solubility of Aripiprazole by Multicomponent Crystallization. *Crystals* **2021**, *11*, 343. [CrossRef]
4. Acebedo-Martínez, F.J.; Alarcón-Payer, C.; Rodríguez-Domingo, L.; Domínguez-Martín, A.; Gómez-Morales, J.; Choquesillo-Lazarte, D. Furosemide/Non-Steroidal Anti-Inflammatory Drug–Drug Pharmaceutical Solids: Novel Opportunities in Drug Formulation. *Crystals* **2021**, *11*, 1339. [CrossRef]
5. Saikia, B.; Seidel-Morgenstern, A.; Lorenz, H. Multicomponent Materials to Improve Solubility: Eutectics of Drug Aminoglutethimide. *Crystals* **2022**, *12*, 40. [CrossRef]
6. Benito, M.; Barceló-Oliver, M.; Frontera, A.; Molins, E. Oxalic Acid, a Versatile Coformer for Multicomponent Forms with 9-Ethyladenine. *Crystals* **2022**, *12*, 89. [CrossRef]
7. Castiñeiras, A.; Frontera, A.; García-Santos, I.; González-Pérez, J.M.; Niclós-Gutiérrez, J.; Torres-Iglesias, R. Multicomponent Solids of DL-2-Hydroxy-2-phenylacetic Acid and Pyridinecarboxamides. *Crystals* **2022**, *12*, 142. [CrossRef]
8. Baraldi, L.; Fornasari, L.; Bassanetti, I.; Amadei, F.; Bacchi, A.; Marchiò, L. Salification Controls the In-Vitro Release of Theophylline. *Crystals* **2022**, *12*, 201. [CrossRef]
9. Verdugo-Escamilla, C.; Alarcón-Payer, C.; Acebedo-Martínez, F.J.; Fernández-Penas, R.; Domínguez-Martín, A.; Choquesillo-Lazarte, D. Lidocaine Pharmaceutical Multicomponent Forms: A Story about the Role of Chloride Ions on Their Stability. *Crystals* **2022**, *12*, 798. [CrossRef]
10. Acebedo-Martínez, F.J.; Alarcón-Payer, C.; Barrales-Ruiz, H.M.; Niclós-Gutiérrez, J.; Domínguez-Martín, A.; Choquesillo-Lazarte, D. Towards the Development of Novel Diclofenac Multicomponent Pharmaceutical Solids. *Crystals* **2022**, *12*, 1038. [CrossRef]

Disclaimer/Publisher's Note: The statements, opinions and data contained in all publications are solely those of the individual author(s) and contributor(s) and not of MDPI and/or the editor(s). MDPI and/or the editor(s) disclaim responsibility for any injury to people or property resulting from any ideas, methods, instructions or products referred to in the content.

Review

Recent Advances in the Application of Characterization Techniques for Studying Physical Stability of Amorphous Pharmaceutical Solids

Yanan Wang [1,†], Yong Wang [1,†], Jin Cheng [1], Haibiao Chen [2], Jia Xu [1], Ziying Liu [1], Qin Shi [1,*] and Chen Zhang [2,*]

1. School of Pharmacy, Jiangsu Vocational College of Medicine, Yancheng 224005, China; 11997@jsmc.edu.cn (Y.W.); 11956@jsmc.edu.cn (Y.W.); 11422@jsmc.edu.cn (J.C.); xujia@jsmc.edu.cn (J.X.); 11707@jsmc.edu.cn (Z.L.)
2. Institute of Marine Biomedicine, Shenzhen Polytechnic, Shenzhen 518055, China; chenhb@szpt.edu.cn
* Correspondence: 12127@jsmc.edu.cn (Q.S.); zhangchen@szpt.edu.cn (C.Z.)
† Yanan Wang and Yong Wang contributed equally to this work.

Abstract: The amorphous form of a drug usually exhibits higher solubility, faster dissolution rate, and improved oral bioavailability in comparison to its crystalline forms. However, the amorphous forms are thermodynamically unstable and tend to transform into a more stable crystalline form, thus losing their advantages. In order to investigate and suppress the crystallization, it is vital to closely monitor the drug solids during the preparation, storage, and application processes. A list of advanced techniques—including optical microscopy, surface grating decay, solid-state nuclear magnetic resonance, broadband dielectric spectroscopy—have been applied to characterize the physicochemical properties of amorphous pharmaceutical solids, to provide in-depth understanding on the crystallization mechanism. This review briefly summarizes these characterization techniques and highlights their recent advances, so as to provide an up-to-date reference to the available tools in the development of amorphous drugs.

Keywords: amorphous drug; characterization methods; crystallization; physical stability; molecular mobility

1. Introduction

Oral administration is a preferred route for drug delivery in many cases because of it is easy, convenient, and safe for most patients. However, many new drug candidates with low solubility fail to reach the target bioavailability via oral administration [1]. One effective approach to enhance the solubility of poorly water-soluble drugs is to keep the drugs in their amorphous form [2–4]. Compared to crystals, amorphous solids are inherently in the higher-energy state and thus their dissolution is more energy favorable [5]. However, the higher-energy of amorphous solids can also cause phase instability and drive them towards crystallization, causing the loss of their advantages being amorphous.

A common approach to improve the stability of the amorphous drugs is to disperse them into polymeric materials to yield kinetically stable amorphous solid dispersions [6–16]. To obtain a better insight into the physical stability of amorphous pharmaceutical solids, it is of special importance to investigate the key physicochemical properties governing the crystallization and phase separation. In the past few decades, key thermodynamic and kinetic properties of amorphous solids have been extensively studied, including configurational entropy, structural relaxation, secondary relaxation, surface molecular diffusion, etc. [17–20]. These thermodynamic and kinetic properties have attracted wide attentions because they are considered important parameters for predicting the physical stability of amorphous pharmaceutical solids.

Unique properties of crystalline and amorphous solids have been extensively investigated by several established and emerging techniques including polarized light microscope

combined with a hot stage, surface grating decay, thermal analysis, broadband dielectric spectroscopy, and solid-state nuclear magnetic resonance (NMR). In this article, we will review the recent advances and highlight the applications of these characterization techniques. Moreover, we will also briefly discuss the limitations, challenges, and future development trends of these characterization techniques, and the aim is to provide a reference of current tools for developing robust amorphous pharmaceutical formulations.

2. Combination of Polarized Light Microscope and Hot Stage

Polarized light microscope is one of the most widely used techniques in characterizing nucleation [21–23], crystal growth [24,25], polymorphic transition [26], and phase separation [27] of amorphous pharmaceutical solids. In recent studies, polarized light microscopy was often combined with a hot stage to extend its experimental temperature range and achieve the goal of precise temperature control. Prior to investigating the nucleation and crystal growth processes, amorphous drug was firstly prepared by the melting-quenching method using a hot stage for precisely controlling the temperature. The drugs prepared by the melt-quenching method are confirmed to be amorphous by the absence of birefringence under the polarized light microscope.

Cai and co-workers systemically investigated the crystal growth behaviors of a classical antifungal drug griseofulvin as a function of temperature by a polarized light microscope equipped with a hot stage [24]. As shown in Figure 1, growth morphologies of griseofulvin crystals exhibit strong temperature dependence, producing faceted coarse crystals near the melting point (T_m), fiber-like fine crystal near the glass transition temperature (T_g), and even finer compact spherulites below T_g [24]. In addition, the velocity of crystal growth of griseofulvin could also be measured by recording the advancing growth front of the crystal into the supercooled liquid or glass as a function of time [24]. In the supercooled liquid, the rate of crystal growth decreases with the temperature decreasing near T_m while it increases with the temperature further decreasing near T_g. This bell-shaped curve of crystal growth rate vs. temperature is mainly attributed to the competition between the negative dependence of the thermodynamic driving force and the positive dependence of the bulk molecular diffusion on the temperature. For some small-molecule drugs, one fast crystal growth mode termed as glass-to-crystal (GC) growth, is activated as the temperature decreases near or below T_g, with a rate orders of magnitude faster than those predicted by bulk diffusion controlled modes [24,28]. Several models have been proposed to explain the GC growth; however, its mechanism remain imperfectly understood. In a very recent model, voids and free surface is proposed to be continuously created by fracture, leading the fast GC growth by taking advantage of the fast surface mobility.

For the crystal growth in drug–polymer binary systems, one interesting phenomenon, polymer enrichment as a function of temperature and polymer concentration, could be observed at the advancing growth front of crystal using polarized light microscopy combined with a hot stage [29,30]. Visual observations can directly provide a reasonable explanation for the fact that the increase of global molecular mobility alone is insufficient to account for the accelerating effects of a low-T_g polymer—e.g., poly (ethylene oxide) (PEO)—on the crystal growth of small-molecule drug griseofulvin and indomethacin [31,32]. In addition to the high global molecular mobility, PEO enrichment at the growth front could also accelerate the mass transport of drug molecules entering the crystalline phase by the high segmental mobility of PEO [29,30]. Moreover, polymer enrichment could also be one of the key factors rendering the selective effect of polymer on the crystal growth of different drug polymorphs [30]. In a very recent study, Zhang et al. found that the extent of the accelerating effect of PEO on the crystal growth of indomethacin polymorphs is very consistent with the concentration of PEO enrichment at the crystal–liquid interface [30]. They proposed that distinct drug–polymer distribution at the growth front of indomethacin polymorphs strongly affects the mass transport of drug molecules and the energy barrier, thus leading to the selective accelerating effects of PEO on crystal growth of indomethacin polymorphs [30].

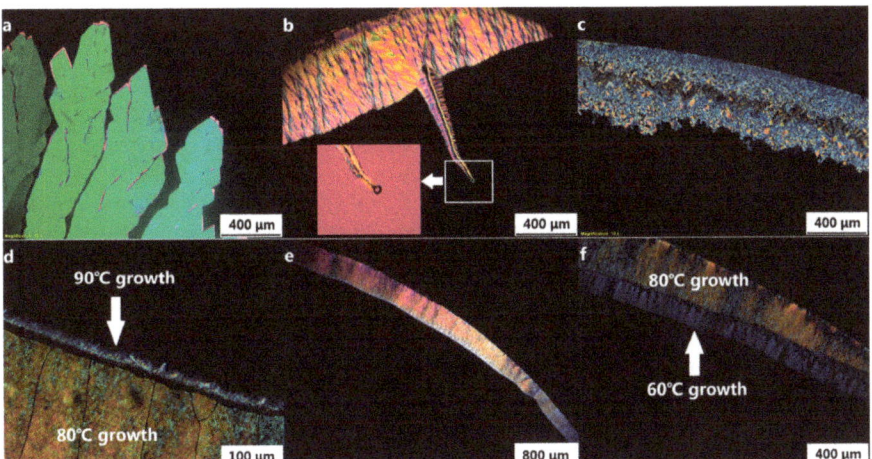

Figure 1. Bulk crystal growth morphologies of GSF as a function of temperature from 210 °C to 60 °C, (**a**) 210 °C, (**b**) 130 °C, (**c**) 100 °C, (**d**) 90 °C, (**e**) 80 °C, and (**f**) 60 °C. Adapted from the [24] with the permission. Copyright © 2016 American Chemical Society.

Apart from studying crystal growth behaviors, polarized light microscope combined with hot stage can be used to explore the nucleation of amorphous pharmaceutical solids [21–23]. According to the different kinetics of nucleation and crystal growth, studies on drug nucleation can be mainly defined as one-stage method and two-stage method [21–23]. One-stage method can be used to determine the number of nucleation events per unit volume for the drug systems showing both fast nucleation and fast crystal growth behaviors. Herein, individual nucleation is allowed to grow to the observable size for some time t_0 in the supercooled liquid. The size and growth rates of the crystal are measured by using the combined technology of polarized light microscope and hot stage. On the basis of the crystal size (radius r) and growth rate (u), the birth time t of individual nucleus as a function of time can be calculated as

$$t = t_0 - r/u$$

In general, drug nucleation follows a steady rate after an induction period, followed by a slower rate due to the decreased available liquid volume for nucleation.

If a drug crystal grows relatively slow, one-stage method is not suitable, and two-stage method needs to be applied for this situation. Unlike one-stage method, two-stage method is briefly summarized as a two-step process consisting of a nucleation step at a relatively low temperature and then quickly switch to an elevated temperature to allow the nuclei grow to visible dimensions under a polarized microscope. The temperature selected for crystal growth is required to ensure the rapid growth of nuclei but meanwhile prevent the formation of new nuclei.

Huang et al. compared the crystal nucleation rates of four polyalcohols exhibiting the similar kinetics of crystal growth on the T/T_g scale [21]. On the same scale of T/T_g, nucleation rates of these four polyalcohols are vastly different, indicating the fundamentally different mechanisms of nucleation [21]. In a recent study, Yao et al. compared the inhibitory effects of polymer on the crystal nucleation and growth via polarized light microscope combined with a hot stage [22]. Interestingly, the inhibitory effects of polymer on the crystal nucleation rates are similar to those on the crystal growth [22]. At a given temperature, the ratios between the rates of nucleation and growth are nearly identical, and are independent of the concentration and molecular weight of the polymer [22]. Moreover, in a very recent study, Zhang et al. found that the accelerating effects of low-T_g polymer (PEO) on crystal nucleation and growth of fluconazole are also approximately the same [23]. Based on these studies, both the crystal nucleation and growth were proposed to be molecular mobility-limited processes [22,23]. Herein, dissolved polymer in the amorphous matrix acts as a mobility modifier, imposing similar degrees of inhibitory and accelerating effects on the crystal nucleation and growth [22,23].

One important nucleation phenomenon, termed as cross-nucleation, could also be observed under a polarized light microscope, in which another polymorph nucleates on the early nucleating polymorph [33]. Compared to the early nucleating polymorph, the newly nucleated polymorph could be less or more thermodynamically stable [34]. This interesting nucleation behavior is quite different from the classical Ostwald's law of stages, would lead to the ineffectiveness of the seeding method for obtaining the required polymorph. The newly nucleated polymorph always exhibits a faster or same crystal growth rate as the initial polymorph. If the frequency of cross-nucleation is sufficiently high, thesurface of the early nucleating polymorph will eventually be occupied by the cross-nuclei of newly nucleated polymorph.

Polarized light microscope combined with hot stage can also be used to obtain high-quality single crystals [35,36]. For instance, high-quality single crystals of the metastable form II of griseofulvin, was obtained from the melt under a polarized light microscope equipped with a hot stage [35]. Large and faceted single crystals of griseofulvin form II were observed after a rapid growth at 200 °C in the presence of 10% w/w PEO, which is reported to effectively accelerate the crystal growth [35]. Single-crystal X-ray diffraction analysis as a function of temperature revealed that griseofulvin form II exhibited an anomalously large coefficient of thermal expansion [35]. In a recent study, a creative strategy of cultivating single crystal was developed for rapidly obtaining the desired polymorph from the melt microdroplets near T_m [35]. Herein, a hot stage was used to control the temperature near T_m, at which the secondary nucleation was effectively inhibited to avoid the formation of polycrystals. Meanwhile, polarized microscope was used to monitor the growth of single crystals until a proper size was reached. In addition, polarized light microscope combined with hot stage can also be applied into the downstream processing of the preparation of amorphous pharmaceutical formulations [37].Yang et al. used this combined technique to analyze the microstructure and state of the samples at various temperature, facilitating the determination of temperature range for amorphous drug formulations during the hot melt extrusion process [37].

3. Surface Grating Decay and Surface Diffusion

Surface mobility can considerably affect processes including nucleation, crystal growth, catalysis, and sintering. The high mobility of surface molecules originates from the special coordination environment, where molecules have fewer neighbors and a greater degree of freedom compared to the molecules in the interior [20,38,39]. For amorphous pharmaceutical solids, a high surface mobility causes the rapid nucleation and crystal growth at the free surface or interior interface, largely determining the physical stability [20,25,27,40]. Moreover, a high surface mobility could also allow the efficient equilibrium of newly deposited molecules in vapor deposition, facilitating the formation of highly stable glass with exceptionally low energy and high density [41,42].

Surface grating decay method has been widely used to measure the surface diffusion of pure amorphous drug and polymer-doped solid dispersion in the pharmaceutical field [20,43–45]. Herein, surface diffusion represents the lateral movement of molecules at the free surface. As shown in Figure 2, a master grating with gold coating is placed on the surface of drug liquid below T_g to print the surface grating. The master can be detached at a lower temperature and a pharmaceutical glass with a corrugated surface is produced. The smoothing of the surface grating is followed with an atomic force microscope or optical microscope under nitrogen atmosphere. The amplitude of surface grating could be obtained by the Fourier transformation of the height profile of each scan line. In general, grating amplitude h decreases exponentially following $h = h_0 \exp(-Kt)$ in the supercooled liquid state. For comparison, h decays slightly deviating from exponentially and could be described as $h = h_0 \exp[-(Kt)^\beta]$ in the glassy state, one phenomenon mainly attributed to the glass aging. Here, K represents the decay constant and β is slightly smaller than 1. According to the Mullin's model, the decay constant K can be described as a combination of individual processes including viscous flow (F), evaporation condensation (A and A'), bulk diffusion (C), and surface diffusion (B) in the following equation.

$$K = Fq + Aq^2 + Dq^3 + Bq^4$$

$$q = \frac{2\pi}{\lambda} \quad F = \frac{\gamma}{2\eta} \quad A = \frac{p_0 \gamma \Omega^2}{(2\pi m)^{0.5}(kT)^{1.5}} \tag{1}$$

$$D = A' + C = \frac{\rho_0 D_G \gamma \Omega^2}{kT} + \frac{D_v \gamma \Omega}{kT} \quad B = \frac{D_s \gamma \Omega^2 v}{kT}. \tag{2}$$

where γ represents the surface free energy, η represents the viscosity, p_0 represents the equilibrium vapor pressure, Ω represents the molecular volume, m represents the molecular mass, ρ_0 represents the equilibrium vapor density, D_G represents the diffusion coefficient of evaporated molecules, and v represents the number of molecules per unit area of surface. D_v and D_s is the coefficient of bulk diffusion and surface diffusion, respectively.

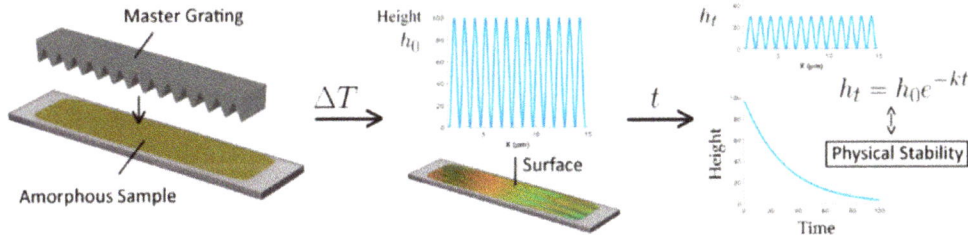

Figure 2. The experimental scheme of surface grating decay for surface diffusion measurement. Adapted from [43] with the permission. Copyright © 2020 American Chemical Society.

In 2011, Zhu et al. investigated the surface diffusion of a small-molecular drug indomethacin by the method of surface grating decay for the first time [44]. Surface evolution of amorphous indomethacin is mainly controlled by the viscous flow at high temperature while the mechanism of surface evolution changes to the surface diffusion with the temperature decreasing to near and below T_g [20,43–45]. Compared to the bulk diffusion, surface diffusion of amorphous drugs can be orders of magnitude faster [44–46]. Moreover, as shown in Figure 3, surface diffusion coefficient D_s has been shown to be roughly proportional to the velocity of surface crystal growth u_s ($u_s \sim D_s^{0.87}$), indicating the controlling role of D_s in the process of surface crystal growth [45].

Recent studies showed that surface diffusion of molecular glasses can be strongly affected by several factors, including strength of molecular interaction [47,48], molecular size [49], and the addition of polymer [43]. With the increase in the strength of molecular interaction and molecular size, surface diffusion of amorphous solids exhibits a tendency to slow down [47–51]. For instance, Chen et al. found that the surface diffusion of polyalcohol glasses showing extensive hydrogen bonding is much slower than that of the molecular glasses of comparable size but with no or limited hydrogen bonds [48]. They proposed that the inhibition of surface diffusion in systems containing extensive hydrogen bonding interactions is mainly attributed to the abundance of hydrogen bonds near the surface [48]. As a result, the loss of nearest neighbors could not induce a proportional decrease in the kinetic barrier of surface diffusion [48]. Surface diffusions of posaconazole and itraconazole were also investigated by surface grating decay method, and the diffusion rates of these two rod-like molecules are much lower than those of quasi-spherical molecules of similar volume, a result of the deep penetration of rod-like molecules in the bulk where molecular mobility is slow [52]. In amorphous systems without extensive hydrogen bonds, surface diffusion coefficients of molecular glasses were proposed to decrease with an increase in their penetration depth [52]. In addition, Mokshin et al. proposed that surface diffusion coefficient D_s is directly related to the kinetic coefficient (also termed as attachment coefficient) of crystallized molecular glasses [53]. In a very recent study, Bannow et al. investigated the effects of a commercial polymeric excipient Soluplus on the surface diffusion of amorphous indomethacin [43]. The addition of low-concentration Soluplus significantly slowed down the surface diffusion of indomethacin [43]. Further increase in the concentration of Soluplus would lead to turnover, where the increasing inhibitory effect of Soluplus on the surface diffusion with the Soluplus concentration increasing becomes less pronounced [43]. Moreover, the decrease in surface diffusion of amorphous indomethacin by doping low content Soluplus correlates well with the enhanced physical stability [43].

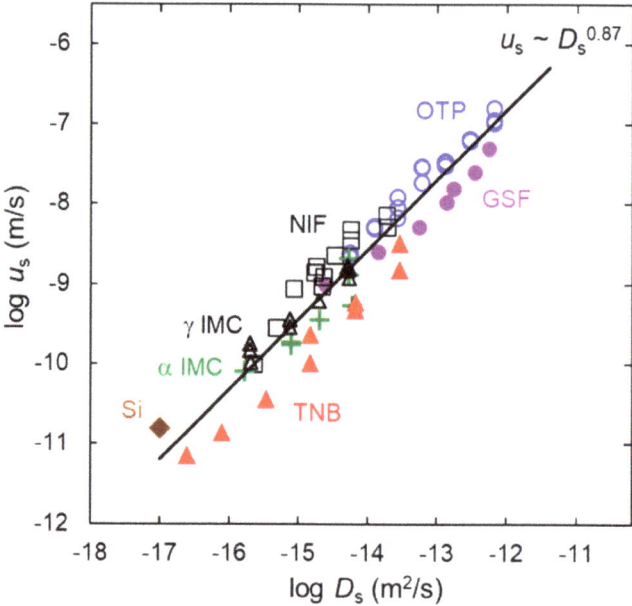

Figure 3. Surface crystal growth rate u_s, plotted against surface diffusion coefficients D_s for reported glasses and silicon. Adapted from [45] with the permission. Copyright © 2017 American Chemical Society.

4. Solid-State Nuclear Magnetic Resonance (NMR)

Solid-state nuclear magnetic resonance (SS-NMR) has also been introduced into the pharmaceutical field for studying the dynamics and phase composition of amorphous solid dispersions [54,55]. In addition, crystalline and amorphous drug generally exhibit different spectra of SS-NMR [37,56]. Compared to the crystalline form, ^{13}C peaks of amorphous drug are much broader, which is mainly attributed to the disordered molecular packing [56]. SS-NMR spectroscopy could provide diverse and critical information of complex amorphous dispersions from the atomic level, which is barely realized by other existing methods [54]. One main application of SS-NMR is to study the site-specific molecular mobility of amorphous solids by measuring the spin–lattice relaxation times and their spin–spin relaxation [57]. Herein, the spin–lattice relaxation times of rare nuclei consist of the static and rotating frame form (T_1 and $T_{1\rho}$) [57]. T_1 and $T_{1\rho}$ relaxations (e.g., ^{13}C) could provide pure dynamic information including the most kinds of local motions [57]. Moreover, relaxations representing the primary and secondary molecular motions can be identified from the spin–lattice relaxation in the SS-NMR [58]. Herein, T_1 relaxation generally represents the secondary molecular relaxations due to its sensitivity towards the local and rapid motions [58]. $T_{1\rho}$ relaxation could be affected by the slower molecular motions of amorphous solid, which is associated with the primary relaxation [58]. For instance, in the case of pharmaceutical polymer methylcellulose and polyvinylpyrrolidone, $T_{1\rho}$ relaxation originates from the motions of polymer backbone. For comparison, T_1 relaxations come from the local dynamics originated from the side chain motions.

High resolution ^{13}C SS-NMR could be used to investigate the hydrogen bonding interactions of amorphous pharmaceutical solids [59]. With the aid of SS-NMR, Yuan et al. quantitatively studied the hydrogen bonding interactions in amorphous indomethacin with and without the presence of poly (vinylpyrrolidone) (PVP) and poly(vinylpyrrolidone-co-vinyl acetate) (PVPVA) [59]. For the pure amorphous indomethacin system, ^{13}C SS-NMR revealed that its hydrogen bonding interaction consists of three main types including disordered carboxylic acid chains, carboxylic acid cyclic dimer, and the carboxylic acid hydrogen bonded to the amide carbonyl (Figure 4) [59]. This self-interaction of indomethacin could be disrupted once indomethacin formed a solid dispersion with the addition of polymer [59]. The extent of drug–polymer hydrogen bonding interactions increase with the increase in polymer concentration. For comparison, the carboxylic dimers between two indomethacin molecules could not be observed any more as the polymer concentration increased to 50% w/w. In a recent study, Sarpal et al. used the SS-NMR technique to compare the molecular interactions in amorphous ketoconazole (KET), KET binary dispersions, and KET ternary dispersions [60]. A detailed ^{13}C SS-NMR deconvolution study showed that binary KET and poly (acrylic acid) (PAA) system exhibits higher prevalence of ionic and hydrogen bonds in comparison with its ternary system containing hydroxypropyl methyl cellulose (HPMC) [60].

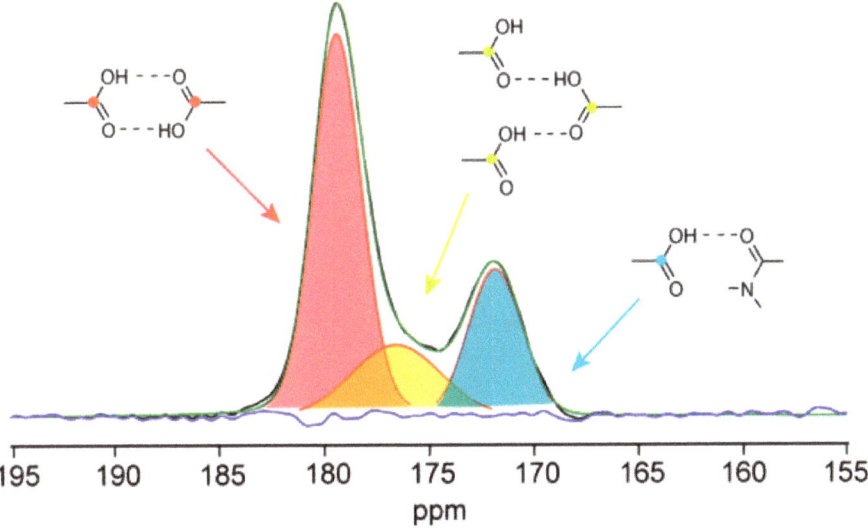

Figure 4. CPMAS ^{13}C spectrum of the carboxylic acid of amorphous indomethacin. Simulated peaks are shown in red, yellow, and blue to illustrate the various hydrogen bonds. Adapted from the [59] with the permission. Copyright © 2015 American Chemical Society.

In recent studies, two-dimensional (2D) ss-NMR was also developed to identify the type and strength of various drug–polymer interactions in ASDs with enhanced sensitivity and resolution [61,62]. Lu et al. used 2D ^{1}H-^{19}F ss-NMR to investigate the molecular interaction between the difluorophenyl group of posaconazole and the hydroxyl group of hydroxypropyl methylcellulose acetate succinate (HPMCAS) in the ASD. For hydrogen bond patterns, they proposed that the hydroxyl groups of HPMCAS act as acceptors while the fluorine or difluorophenyl rings of posaconazole act as donors [62]. Moreover, a ^{19}F-^{13}C rotational echo and double resonance technique was used to measure the atomic distance, and it revealed the close proximity between ^{13}C of the hydroxyl group and ^{19}F of posaconazole at 4.3 Å [62].

Moreover, the miscibility between drug and polymer can also be investigated using the T_1 and $T_{1\rho}$ relaxation times obtained from ^{1}H SS-NMR [59,63–67]. At present, measuring the T_g using differential scanning calorimetry (DSC) is the most widely used approach for determining the miscibility of amorphous solids. Generally, a single T_g between those of drug and polymer indicates a miscible system, while two separated T_gs suggest a phase-separated system. However, T_g is sometimes not a reliable indicator of system miscibility, as some studies reported that phase-separated system could exhibit a single T_g and vice versa [63]. Moreover, the miscibility assessed by T_g generally has a detection limit of 20–30 nm, meaning that a two-phase system with smaller domain size would be indistinguishable by DSC. For comparison, SS-NMR is suggested to be a more accurate technique, which could measure the drug–polymer miscibility for small domain sizes.

For the miscibility measurement using SS-NMR, length scale of ^{1}H spin diffusion is important and could be calculated by

$$\langle L \rangle = \sqrt{6Dt}$$

Herein, t represents the relaxation time, D represents the spin diffusion coefficient and is typically assumed as 10^{-12} cm^2/s for the organic solids. The length scale of spin diffusion is ca. 20–50 nm for a typical T_1 value as 1–5 s. For a typical $T_{1\rho}$ as 5–50 ms, the length scale is ca. 2–5 nm. It is expected that three scenarios might occur depending on the domain size. A common T_1 and $T_{1\rho}$ values could be obtained for drug and polymer for the domain size smaller than 2–5 nm. If the domain size is 5–20 nm, T_1 values of drug and polymer are the same while $T_{1\rho}$ values are different. If both the T_1 and $T_{1\rho}$ values of drug and polymer are different, the domain size is larger than 20–50 nm. Pham et al. used the SS-NMR cross-polarization hetero-nuclear correlation technique to investigate the spin diffusion effects of the amorphous pharmaceutical solids, facilitating the detection of the drug–polymer molecular interaction and phase separation [64]. Litvinov et al. investigated the phase behavior of miconazole-poly(ethylene glycol)-g-vinyl alcohol (PEG-g-PVA) solid dispersions by a combination of modulated DSC, XRPD, and SS-NMR [63]. In their work, miconazole (10% w/w) was identified to form the amorphous nanocluster with ~1.6 nm average cluster size in the amorphous matrix of PEG-g-PVA, indicating the miscibility at the molecular level [65].

Sarpal et al. compared the phase homogeneity of felodipine ASDs doped with PVP, PVPVA, and poly(vinylacetate) (PVAc) by measuring the 1H T1 and T1ρ of drug and polymer [66]. Better compositional homogeneity was observed in felodipine-PVP and felodipine-PVPVA ASDs compared to felodipine-PVAc ASD. ^{13}C ss-NMR was also used to investigate the strength and extent of drug–polymer hydrogen bonding interactions, and it revealed a ranking of PVP > PVPVA > PVAc [66]. The results also suggested that hydrogen bonding interaction in ASDs could impact system phase homogeneity [66]. Recently, an ^{1}H double-filtering SS-NMR technique with ^{1}H spin diffusion and ^{13}C detection was developed to provide the highest-resolution quantification of molecular miscibility and homogeneity of amorphous solid dispersions [68]. However, spectrum acquisition typically requires one week for obtaining sufficient signal-to-noise ratio, therefore, low-field and benchtop NMR techniques are to be developed to shorten the acquisition times, which would make SS-NMR a useful technique for capturing structural evolution that occurs within a short time period [69].

5. Broadband Dielectric Spectroscopy

Recent studies showed that molecular mobility is probably the most relevant factor for reliably predicting the crystallization behavior of amorphous pharmaceutical solids [19,70]. However, it should be noted that amorphous pharmaceutical solids exhibit complex molecular structures accompanied with various configurational topologies and a variety of molecular interactions. Consequently, molecular mobility of amorphous pharmaceutical solids is rather complex in general, and it could be reflected in various relaxation processes originated from different natures. As a result, establishing proper correlations between molecular mobility and physical stability of amorphous pharmaceutical solids is quite challenging.

In recent studies, a variety of strategies have been exploited to investigate the molecular mobility in glassy and supercooled liquid state—including light scattering, mechanical spectroscopy, temperature modulated differential scanning calorimetry (TMDSC), and nuclear magnetic resonance (NMR). Among these approaches, broadband dielectric spectroscopy is able to explore different relaxation processes and has been demonstrated to distinguish the global and local molecular motions. The measurement of broadband dielectric spectroscopy could be performed over an extremely wide frequency range from mHZ to THz, and concurrently in wide temperature and pressure ranges.

In an excellent review, Paluch and coworkers give a detailed introduction for the equipment principle and basic parameters of broadband dielectric spectroscopy [19]. In brief, dielectric measurement is based on the interactions between the electric dipole moment and the charges of the sample when an external electric field is applied. The investigated pharmaceutical material is firstly placed in a sample capacitor. Herein, a generator (e.g., sine wave generator) could apply an alternating voltage $U^*(\omega)$ to the capacitor. Consequently, the external alternating electric field $E(\omega)$ is generated on the sample capacitor. The complex impedance $Z^*(\omega)$ of the samples is determined by the impedance analyzer through measuring the sample capacitor complex voltage and the current. From the basic quantity of complex electrical impedance $Z^*(\omega)$ measured by BDS, other complex quantities such as complex dielectric permittivity $\varepsilon^*(\omega)$ could be derived.

In the field of pharmaceuticals, the Havriliak–Negami (HN) functions plus dc-conductivity term are usually used for the analysis of measured isothermal dielectric spectrum.

$$\varepsilon^*(\omega) = \varepsilon'(\omega) - i\varepsilon''(\omega) = \varepsilon_\infty + \frac{\varepsilon_s - \varepsilon_\infty}{(1 + (i\omega\tau_{HN})^\alpha)^\beta} + \frac{\sigma_U}{i\omega\varepsilon_0} \quad (3)$$

Here, ε' and ε'' respectively represent the real and imaginary parts of the complex dielectric permittivity. ε_∞ represents the high-frequency limit permittivity and ε_s represents the static dielectric constant. α, β are shape parameters of the dielectric peaks and they respectively represent the asymmetry and width. $\sigma_{dc}/i\varepsilon_0\omega$ represents the conductivity component, where σ_{dc} represents the level of dc-conductivity and ε_0 represents the vacuum permittivity.

α-relaxation reflects the reorientations of entire molecules for the low-molecular weight materials. In the case of polymer, α-relaxation, also termed as segmental relaxation, is related to some segmental motions for the polymer chain, which would lead the conformational change. α-relaxation time (τ_α), representing the relaxation time at a maximum loss (τ_{max}). The value of τ_α was calculated by the following equation on the basis of the parameters obtained by the HN function.

$$\tau_\alpha = \tau_{max} = \tau_{HN}\left[\sin\left(\frac{\pi\alpha}{2+2\beta}\right)\right]^{-1/\alpha}\left[\sin\left(\frac{\pi\alpha\beta}{2+2\beta}\right)\right]^{1/\alpha} \quad (4)$$

Empirical Vogel–Fulcher–Tamman (VFT) equation is most widely used for describing the temperature dependence of τ_α in the supercooled liquid state.

$$\tau_\alpha = \tau_0 \exp\left(\frac{DT_0}{(T-T_0)}\right) \quad (5)$$

In this equation, τ_α represents the α-relaxation time and τ_0 is the relaxation time of the unrestricted material. D represents the strength parameter for a measure of fragility and T_0 represents the zero mobility temperature. Moreover, the dependence of τ_α in the glassy state could be also predicted by the extended VFT equation.

Local motions have either intra- or intermolecular origins, could be reflected in different secondary relaxations. For the secondary relaxation times, their temperature dependence in the glassy state can be commonly fitted to the Arrhenius equations

$$\tau = \tau_0 \exp\left(\frac{\triangle E}{RT}\right) \quad (6)$$

Herein, ΔE represents the activation energy and R represents the universal gas constant.

One important goal is to reveal the dominant type of molecular mobility responsible for physical stability, which has been fervently debated for the past decades. Some argue that global relaxation is responsible for the physical stability while other propose that secondary relaxation is the key controlling factor of physical stability. Kothari et al. investigated the molecular mobility and crystallization kinetics of amorphous drug griseofulvin and nifedipine [71]. In their work, the crystallization kinetics of these amorphous drugs is monitored by powder X-ray diffraction technique (PXRD) above T_g while by PXRD technique equipped with synchrotron X-ray source below T_g [71]. They found that physical stability both in the glassy state and supercooled liquid is strongly related to the α-relaxation rather than the secondary relaxations [71]. Similar strong correlations between α-relaxation and physical stability could also been observed in several other pharmaceuticals including itraconazole [72], trehalose [73], celecoxib [74].

In addition to the pure drug system, α-relaxation has also been reported to play the controlling role for the physical stability in the polymer-based amorphous solid dispersions (ASD) [75–77]. Suryanarayanan and co-workers proposed that the formation of nifedipine-polymer hydrogen bonding interactions could translate to a high resistance to the crystallization by reducing the global mobility, as evidenced by longer system α-relaxation time [76]. Further study revealed that α-relaxation times of nifedipine ASD increase linearly with the polymer concentration increasing [75]. In addition, the established relationship between α-relaxation time and crystallization kinetics of nifedipine ASD doped with a low content of polymer could be used as a reliable predictor for the crystallization of nifedipine ASD containing a higher content of polymer [75]. The usefulness of this predictive model is well confirmed as the matching of the predictive and experimental results of physical stability [75]. Moreover, Mistry et al. reported that the stronger drug–polymer interactions could lead to a longer delay before the onset of crystallization, indicating the enhanced physical stability [78]. Interestingly, the correlation between α-relaxation and crystallization times is almost unaffected by the formation and strength of drug–polymer molecular interactions [78]. Mohapatra et al. investigated the effects of molecular weight of PVP on the molecular mobility and crystallization of PVP-indomethacin ASDs [77]. SS-NMR revealed that drug–polymer hydrogen bonding interaction is independent of the molecular weight of PVP [77]. For comparison, the dependences of viscosity and molecular mobility on temperature are reasonably similar for indomethacin ASDs containing PVP with various molecular weight [77]. It is concluded that increased viscosity would also translate to reduced global molecular mobility (e.g., α-relaxation times), and thus effectively inhibit the crystallization of ASDs [77].

Recent studies revealed that molecular mobility of amorphous drug and ASD could be enhanced by the addition of water [79,80], glycerol [81], and low-T_g polymer, e.g., poly(ethylene oxide) [31,32]. Mehta et al. proposed that the increased molecular mobility of amorphous system by the sorbed water is attributed to the plasticization effect [79]. This view is strongly supported by the fact that relaxation times of systems containing different water content overlapped on a temperature scaling of T_g/T [79]. Moreover, as shown in Figure 5, a single linear relationship could be observed for the temperature dependence of crystallization t_c in both dry and water-sorbed griseofulvin systems [79]. Herein, t_c represents the time taken for 0.5% of griseofulvin to crystallize. These results indicate that plasticization effect is also the underlying mechanism for the physical stability of amorphous solids [79]. Given that the coupling extent between α-relaxation and crystallization times of ASD remain the same in the presence of low content of water, a predictive model was built by using the sorbed water for studying the crystallization of some slow crystallizing systems [80]. Similarly, Fung et al. demonstrated that glycerol could also act as a plasticizer, which facilitates the development of an accelerated physical stability testing model of ASDs [81]. The success of this predictive model is mainly attributed to the idea that glycerol could effectively accelerate the crystallization without affecting the mobility–crystallization coupling [81].

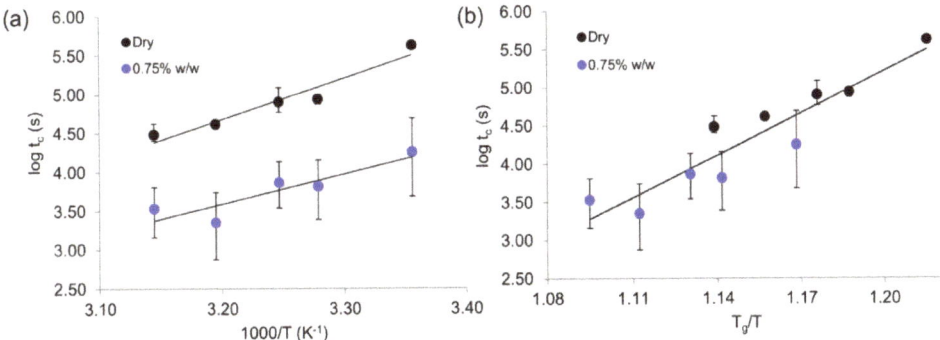

Figure 5. Plot of crystallization times tc as a function of (**a**) inverse temperature (1000/T), and (**b**) T_g/T for the griseofulvin dry powder (black rounds) and powder containing low content water (blue rounds). Adapted from the [79] with the permission. Copyright © 2015 American Chemical Society.

However, some studies found that change in the global molecular mobility characterized by BDS might not be sufficient for explaining the accelerating or inhibitory effects in the crystallization of amorphous drugs [31,32,82]. For instance, PEO plasticizes amorphous drug systems and increases their global mobility from the liquid dynamic perspective, which is evidenced by the overlapping of α-relaxation time at the scale of T_g/T [31,32]. However, from the perspective of crystallization kinetics, the accelerating growth rates of griseofulvin crystals cannot be simply explained by the increased molecular mobility, as evidenced by the fact that growth rates of pure griseofulvin could not overlap with that of griseofulvin containing low content PEO at the scale of T_g/T [31,32]. Moreover, the increase in global mobility (i.e., the decrease in α-relaxation time) also has difficulty in explaining the selective accelerating effects of PEO on the crystallization of different polymorphs of a drug [32]. In addition, the coupling between crystallization kinetics and α-relaxation times could also be affected by the addition of some excipients, indicating that factors other than global mobility for governing the physical stability [82]. In addition, an increasing number of studies proposed that local mobility rather than global mobility could be the major factor for influencing the physical stability in the glassy state [83,84]. For instance, in the case of glassy celecoxib and indomethacin, strong correlations could be observed between the physical instability and Johari–Goldstein (β) relaxation time rather than the α-relaxation time [84]. This intermolecular secondary β-relaxation is proposed to be a precursor of the global molecular mobility, which indicates that these small-angle reorientations would lead the cooperative α-relaxation process.

Analogous to the polymer-based ASDs, BDS has also been performed to study the molecular dynamics in coamorphous formulations [85–89]. Knapik et al. investigated the molecular mobility of ezetimib-indapamide coamorphous systems and its correlations with physical stability [85]. With the increase of indapamid content, physical stability of this binary coamorphous system was progressively enhanced, as evidenced by the longer α-relaxation time and smaller fragility [85]. In addition, T_gs of ezetimib-indapamid coamorphous systems rose with the increasing content of indapamid in accordance with the prediction by Gordon-Taylor equations [85]. They proposed that antiplasticizing effect exerted by indapamid is the main mechanism for improving system physical stability [85]. Fung et al. explored the effects of organic acid for stabilizing amorphous ketoconazole, a weakly basic active pharmaceutical ingredient (API) [88]. With an increase in the strength of drug–acid molecular interactions, molecular mobility of these coamorphous systems decreases, as evidenced by the longer α-relaxation time [88]. However, in the case of ketoconazole–tartaric acid and ketoconazole–citric acid, the decreased global molecular mobility was not sufficient to explain the enhanced physical stability [88]. They proposed that structural factors would also enhance the physical stabilization of these drug–acid coamorphous systems [88]. Unlike oxalic and succinic acids, each critic acid molecule has three carboxylic acid groups, which are more beneficial to the formation of drug–acid hydrogen bonding interaction [88]. The hydroxyl group in tartaric and citric acid would also act as the donors of hydrogen bonds and further facilitate the formation of stronger drug–acid hydrogen bonding interactions [88].

BDS can also be applied for studying the molecular mobility of amorphous drug under the nanoconfinement effects [90,91]. Knapik et al. investigated the effects of nanoconfinement on the molecular mobility and crystallization of amorphous drug ezetimibe [90]. Amorphous ezetimib would still exhibit the tendency to crystallize in the pores of Aeroperl 300 (~30 nm pore size), while no crystallization occurs once the drug was incorporated in the pores of Neusilin US2 (<5 nm pore size) [90]. As shown in Figure 6, compared to the ezetimib-Aeroperl 300 system, α-relaxation time of ezetimib increases once incorporated into the pores of Neusilin US2 [90]. Moreover, BDS experiments also revealed the distinguishable phases of the loaded ezetimib in these two commercially used porous materials [90]. One is associated with the molecules at the pore surface–liquid interface while the other one is connected to the molecules in the inner of pores [90]. Herein, the dramatic stabilization of amorphous ezetimib in the pores of Neusilin US2 could be attributed to an interplay of three factors [90]. One factor is the decreased global molecular mobility of amorphous ezetimib under the nanoconfinement [90]. The other two factors are mainly attributed to immobilization effects of the pore wall and the smaller pore size in comparison to the critical nucleation size of amorphous ezetimib, respectively [90]. In a recent study, Zhang et al. explored the molecular mobility of amorphous drug griseofulvin and indomethacin in anodic aluminum oxide (AAO) templates as a function of pore size [91]. A typical two-layer model was also observed in the indomethacin/AAO system, as evidenced by two separated T_gs and interfacial polarization relaxation in BDS experiments [91]. For the core–shell two-layer model, shell molecules interacting with the walls of nanopores show the higher T_g, while the core molecules exhibit the fast dynamics with the lower T_g. In the case of griseofulvin/AAO system, fast cooling would lead to a metastable three-layer model, featuring the existence of thermodynamic nonequilibrium interlayer in addition to the core and interfacial layer [91]. For comparison, stable core–shell two-layer model instead of unstable three-layer model was observed for griseofulvin in AAO templates using the slow cooling process (0.5 °C/min) [91].

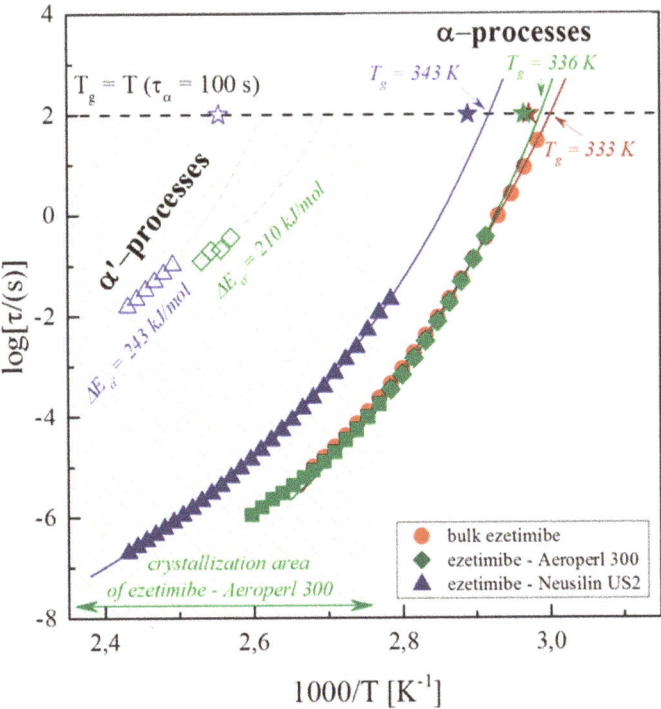

Figure 6. Temperature dependence of α and α' relaxation times determined by using BDS technique for bulk ezetimibe, ezetimibe-Aeroperl 300 and ezetimibe-Neusilin US2 systems. Adapted from [90] with the permission. Copyright © 2016 American Chemical Society.

6. Other Characterization Techniques

In addition to the above-mentioned techniques, several classical characterization techniques have also been widely used in the field of pharmaceuticals for several decades such as FT-IR, Raman spectroscopy, powder X-ray diffraction (PXRD), differential scanning calorimeter (DSC), etc. [92–95]. These classical characterization techniques have found new applications and they are integrated with other techniques [96,97]. For instance, Purohit et al. investigated the miscibility of itraconazole ASDs by the atomic force microscopy technique coupled with nanoscale IR spectroscopy and nanothermal analysis [97]. These combined analytical techniques are proposed to be promising for investigating the phase behaviors of ASDs with high resolution [97]. In the case of PXRD technique, recent studies revealed that it could monitor the extremely low levels of crystallization with enhanced sensitivity by using the synchrotron X-ray source [71].

Moreover, several emerging approaches are also developed for characterizing amorphous pharmaceutical solids including terahertz spectroscopy, X-ray photoelectron spectroscopy, fluorescence lifetime imaging microscopy, etc. [98–100]. Chen et al. systemically investigated the surface enrichment or depletion of components in ASDs by XPS technique [100]. For these spray dried ADSs, surface composition is quite different from those in the bulk [100]. In addition, enrichment or depletion of the drug on the surface of ASDs was found to be strongly dependent on the drug–polymer combination as well as the molecular weight of polymer [100]. In a recent study, an advanced surface characterization platform was developed by combining XPS and time-of-flight secondary ion mass spectrometry [101]. This platform could provide the quantitative measurement of surface composition with high sensitivity and spatial resolution [101].

7. Concluding Remarks

Amorphization of drugs has great value in research and application due to the benefits of improving the solubility, dissolution, and bioavailability of poorly water-soluble drugs. It is of importance to develop characterization methods for studying the physicochemical properties of ASDs, especially for maintaining their physical stability. With a better understanding on the recent development of the characterization methods, it is expected that the commercialization of ASDs with desired pharmaceutical properties would be greatly accelerated.

Author Contributions: Conceptualization, Q.S., C.Z. and Y.W. (Yanan Wang); Writing—original draft preparation, Y.W. (Yanan Wang) and Y.W. (Yong Wang); Writing—review and editing, Q.S., J.C., H.C., J.X. and Z.L.; Funding acquisition, Q.S. and C.Z. All authors have read and agreed to the published version of the manuscript.

Funding: The authors are grateful for financial support of this work by the National Natural Science Foundation of China (nos. 81803452, 21803004), the Natural Science Foundation of Jiangsu Province (no. BK20211114), Shenzhen Science and Technology Innovation Committee (no. JCYJ20200109141808025) and the National Subject Cultivation Project of Jiangsu Vocational College of Medicine (no. 20204304).

Conflicts of Interest: The authors declare no conflict of interest.

References

1. Di, L.; Fish, P.V.; Mano, T. Bridging solubility between drug discovery and development. *Drug Discov. Today* **2012**, *17*, 486–495. [CrossRef] [PubMed]
2. Murdande, S.B.; Pikal, M.J.; Shanker, R.M.; Bogner, R.H. Solubility advantage of amorphous pharmaceuticals: I. A thermodynamic analysis. *J. Pharm. Sci.* **2010**, *99*, 1254–1264. [CrossRef] [PubMed]
3. Yu, L. Amorphous pharmaceutical solids: Preparation, characterization and stabilization. *Adv. Drug Deliver. Rev.* **2001**, *48*, 27–42. [CrossRef]
4. Brough, C.; Williams, R.O. Amorphous solid dispersions and nano-crystal technologies for poorly water-soluble drug delivery. *Int. J. Pharm.* **2013**, *453*, 157–166. [CrossRef] [PubMed]
5. Babu, N.J.; Nangia, A. Solubility advantage of amorphous drugs and pharmaceutical cocrystals. *Cryst. Growth Des.* **2011**, *11*, 2662–2679. [CrossRef]
6. Mooter, G.V.D. The use of amorphous solid dispersions: A formulation strategy to overcome poor solubility and dissolution rate. *Drug Discov. Today Tech.* **2012**, *9*, 79–85. [CrossRef]
7. Yan, H.; Chris, H. Amorphous solid dispersions: Utilization and challenges in drug discovery and development. *J. Pharm. Sci.* **2015**, *104*, 3237–3258.
8. Powell, C.T.; Cai, T.; Hasebe, M.; Gunn, E.M.; Gao, P.; Zhang, G.; Gong, Y.; Yu, L. Low-concentration polymers inhibit and accelerate crystal growth in organic glasses in correlation with segmental mobility. *J. Phys. Chem. B* **2013**, *117*, 10334–10341. [CrossRef]
9. Kestur, U.S.; Taylor, L.S. Role of polymer chemistry in influencing crystal growth rates from amorphous felodipine. *CrystEngComm* **2010**, *12*, 2390–2397. [CrossRef]
10. Cai, T.; Zhu, L.; Yu, L. Crystallization of organic glasses: Effects of polymer additives on bulk and surface crystal growth in amorphous nifedipine. *Pharm. Res.* **2011**, *28*, 2458–2466. [CrossRef]
11. Huang, C.; Powell, C.T.; Sun, Y.; Cai, T.; Yu, L. Effect of low-concentration polymers on crystal growth in molecular glasses: A controlling role for polymer segmental mobility relative to host dynamics. *J. Phys. Chem. B* **2017**, *121*, 1963–1971. [CrossRef]
12. Ilevbare, G.A.; Liu, H.; Edgar, K.J.; Taylor, L.S. Inhibition of solution crystal growth of ritonavir by cellulose polymers—Factors influencing polymer effectiveness. *CrystEngComm* **2012**, *14*, 6503–6514. [CrossRef]
13. Ilevbare, G.A.; Liu, H.; Edgar, K.J.; Taylor, L.S. Maintaining supersaturation in aqueous drug solutions: Impact of different polymers on induction times. *Cryst. Growth Des.* **2013**, *13*, 740–751. [CrossRef]
14. Schram, C.J.; Beaudoin, S.P.; Taylor, L.S. Polymer inhibition of crystal growth by surface poisoning. *Cryst. Growth Des.* **2016**, *16*, 2094–2103. [CrossRef]
15. Li, N.; Taylor, L.S. Tailoring supersaturation from amorphous solid dispersions. *J. Control Release* **2018**, *279*, 114–125. [CrossRef]
16. Taylor, L.S.; Zhang, G.G. Physical chemistry of supersaturated solutions and implications for oral absorption. *Adv. Drug Deliv. Rev.* **2016**, *101*, 122–142. [CrossRef]
17. Zhou, D.; Grant, D.J.; Zhang, G.G.; Law, D.; Schmitt, E.A. A calorimetric investigation of thermodynamic and molecular mobility contributions to the physical stability of two pharmaceutical glasses. *J. Pharm. Sci.* **2010**, *96*, 71–83. [CrossRef]

18. Laitinen, R.; Lobmann, K.; Strachan, C.J.; Grohganz, H.; Rades, T. Emerging trends in the stabilization of amorphous drugs. *Int. J. Pharm.* **2013**, *453*, 65–79. [CrossRef]
19. Grzybowska, K.; Capaccioli, S.; Paluch, M. Recent developments in the experimental investigations of relaxations in pharmaceuticals by dielectric techniques at ambient and elevated pressure. *Adv. Drug Deliv. Rev.* **2016**, *100*, 158–182. [CrossRef]
20. Yu, L. Surface mobility of molecular glasses and its importance in physical stability. *Adv. Drug Deliv. Rev.* **2016**, *100*, 3–9. [CrossRef]
21. Huang, C.; Chen, Z.; Gui, Y.; Shi, C.; Zhang, G.G.; Yu, L. Crystal nucleation rates in glass-forming molecular liquids: D-sorbitol, D-arabitol, D-xylitol, and glycerol. *J. Chem. Phys.* **2018**, *149*, 054503. [CrossRef]
22. Yao, X.; Huang, C.; Benson, E.G.; Shi, C.; Zhang, G.G.; Yu, L. Effect of polymers on crystallization in glass-forming molecular liquids: Equal suppression of nucleation and growth and master curve for prediction. *Cryst. Growth Des.* **2019**, *20*, 237–244. [CrossRef]
23. Zhang, J.; Liu, Z.; Wu, H.; Cai, T. Effect of polymeric excipients on nucleation and crystal growth kinetics of amorphous fluconazole. *Biomater. Sci* **2021**, *9*, 4308–4316. [CrossRef]
24. Shi, Q.; Cai, T. Fast crystal growth of amorphous griseofulvin: Relations between bulk and surface growth modes. *Cryst. Growth Des.* **2016**, *16*, 3279–3286. [CrossRef]
25. Shi, Q.; Tao, J.; Zhang, J.; Su, Y.; Cai, T. Crack- and bubble-induced fast crystal growth of amorphous griseofulvin. *Cryst. Growth Des.* **2020**, *20*, 24–28. [CrossRef]
26. Srirambhatla, V.K.; Guo, R.; Dawson, D.M.; Price, S.L.; Florence, A.J. Reversible, two-step single-crystal to single-crystal phase transitions between desloratadine forms I., II, and III. *Cryst. Growth Des.* **2020**, *20*, 1800–1810. [CrossRef]
27. Shi, Q.; Li, F.; Yeh, S.; Wang, Y.; Xin, J. Physical stability of amorphous pharmaceutical solids: Nucleation, crystal growth, phase separation and effects of the polymers. *Int. J. Pharm.* **2020**, *590*, 119925. [CrossRef]
28. Wang, K.; Sun, C.C. Crystal growth of celecoxib from amorphous state: Polymorphism, growth mechanism, and kinetics. *Cryst. Growth Des.* **2019**, *19*, 3592–3600. [CrossRef]
29. Zhang, J.; Shi, Q.; Tao, J.; Peng, Y.; Cai, T. Impact of polymer enrichment at the crystal-liquid interface on crystallization kinetics of amorphous solid dispersions. *Mol. Pharm.* **2019**, *16*, 1385–1396. [CrossRef]
30. Zhang, J.; Shi, Q.; Guo, M.; Liu, Z.; Cai, T. Melt crystallization of indomethacin polymorphs in the presence of poly(ethylene oxide): Selective enrichment of the polymer at the crystal-liquid interface. *Mol. Pharm.* **2020**, *17*, 2064–2071. [CrossRef]
31. Shi, Q.; Zhang, C.; Su, Y.; Zhang, J.; Zhou, D.; Cai, T. Acceleration of crystal growth of amorphous griseofulvin by low-concentration poly(ethylene oxide): Aspects of crystallization kinetics and molecular mobility. *Mol. Pharm.* **2017**, *14*, 2262–2272. [CrossRef] [PubMed]
32. Shi, Q.; Zhang, J.; Zhang, C.; Jiang, J.; Tao, J.; Zhou, D.; Cai, T. Selective acceleration of crystal growth of indomethacin polymorphs by low-concentration poly(ethylene oxide). *Mol. Pharm.* **2017**, *14*, 4694–4704. [CrossRef] [PubMed]
33. Yu, L. Nucleation of one polymorph by another. *J. Am. Chem. Soc.* **2003**, *125*, 6380–6381. [CrossRef] [PubMed]
34. Chen, S.; Xi, H.; Yu, L. Cross-nucleation between ROY polymorphs. *J. Am. Chem. Soc.* **2005**, *127*, 17439–17444. [CrossRef] [PubMed]
35. Su, Y.; Xu, J.; Shi, Q.; Yu, L.; Cai, T. Polymorphism of griseofulvin: Concomitant crystallization from the melt and a single crystal structure of a metastable polymorph with anomalously large thermal expansion. *Chem. Commun.* **2018**, *54*, 358–361. [CrossRef] [PubMed]
36. Ou, X.; Li, X.; Rong, H.; Yu, L.; Lu, M. A general method for cultivating single crystals from melt microdroplets. *Chem. Commun.* **2020**, *56*, 9950–9953. [CrossRef]
37. Yang, F.; Su, Y.; Zhang, J.; DiNunzio, J.; Leone, A.; Huang, C.; Brown, C.D. Rheology guided rational selection of processing temperature to prepare copovidone/nifedipine amorphous solid dispersions via hot melt extrusion (HME). *Mol. Pharm.* **2016**, *13*, 3494–3505. [CrossRef]
38. Chai, Y.; Salez, T.; Mcgraw, J.D.; Benzaquen, M.; Dalnokiveress, K.; Raphaël, E.; Forrest, J.A. A direct quantitative measure of surface mobility in a glassy polymer. *Science* **2014**, *343*, 994–999. [CrossRef]
39. Yang, Z.; Fujii, Y.; Lee, F.K.; Lam, C.H.; Tsui, O.K. Glass transition dynamics and surface layer mobility in unentangled polystyrene films. *Science* **2010**, *328*, 1676–1679. [CrossRef]
40. Su, Y.; Yu, L.; Cai, T. Enhanced crystal nucleation in glass-forming liquids by tensile fracture in the glassy state. *Cryst. Growth Des.* **2018**, *19*, 40–44. [CrossRef]
41. Swallen, S.F.; Kearns, K.L.; Mapes, M.K.; Kim, Y.S.; Mcmahon, R.J.; Ediger, M.D.; Wu, T.; Yu, L.; Satija, S. Organic glasses with exceptional thermodynamic and kinetic stability. *Science* **2007**, *315*, 353–356. [CrossRef]
42. Brian, C.W.; Zhu, L.; Yu, L. Effect of bulk aging on surface diffusion of glasses. *J. Chem. Phys.* **2014**, *140*, 054509. [CrossRef] [PubMed]
43. Bannow, J.; Karl, M.; Larsen, P.E.; Hwu, E.T.; Rades, T. Direct measurement of lateral molecular diffusivity on the surface of supersaturated amorphous solid dispersions by atomic force microscopy. *Mol. Pharm.* **2020**, *17*, 1715–1722. [CrossRef] [PubMed]
44. Zhu, L.; Brian, C.W.; Swallen, S.F.; Straus, P.T.; Ediger, M.D.; Yu, L. Surface self-diffusion of an organic glass. *Phys. Rev. Lett.* **2011**, *106*, 256103. [CrossRef]

45. Huang, C.; Ruan, S.; Cai, T.; Yu, L. Fast surface diffusion and crystallization of amorphous griseofulvin. *J. Phys. Chem. B* **2017**, *121*, 9463–9468. [CrossRef]
46. Brian, C.W.; Yu, L. Surface self-diffusion of organic glasses. *J. Phys. Chem. A* **2013**, *117*, 13303–13309. [CrossRef] [PubMed]
47. Zhang, W.; Brian, C.W.; Yu, L. Fast surface diffusion of amorphous o-terphenyl and its competition with viscous flow in surface evolution. *J. Phys. Chem. B* **2015**, *119*, 5071–5078. [CrossRef]
48. Chen, Y.; Zhang, W.; Yu, L. Hydrogen bonding slows down surface diffusion of molecular glasses. *J. Phys. Chem. B* **2016**, *120*, 8007–8015. [CrossRef]
49. Zhang, W.; Yu, L. Surface diffusion of polymer glasses. *Macromolecules* **2016**, *49*, 731–735. [CrossRef]
50. Barták, J.; Málek, J.; Bagchi, K.; Ediger, M.D.; Li, Y.; Yu, L. Surface mobility in amorphous selenium and comparison with organic molecular glasses. *J. Chem. Phys.* **2021**, *154*, 074703. [CrossRef]
51. Chen, Y.; Chen, Z.; Tylinski, M.; Ediger, M.D.; Yu, L. Effect of molecular size and hydrogen bonding on three surface-facilitated processes in molecular glasses: Surface diffusion, surface crystal growth, and formation of stable glasses by vapor deposition. *J. Chem. Phys.* **2019**, *150*, 024502. [CrossRef]
52. Li, Y.; Zhang, W.; Bishop, C.; Huang, C.; Ediger, M.D.; Yu, L. Surface diffusion in glasses of rod-like molecules posaconazole and itraconazole: Effect of interfacial molecular alignment and bulk penetration. *Soft Matter* **2020**, *16*, 5062–5070. [CrossRef]
53. Mokshin, A.V.; Galimzyanov, B.N.; Yarullin, D.T. Scaling relations for temperature dependences of the surface self-diffusion coefficient in crystallized molecular glasses. *JETP Lett.* **2019**, *110*, 511–516. [CrossRef]
54. Paudel, A.; Geppi, M.; Mooter, G.V.D. Structural and dynamic properties of amorphous solid dispersions: The role of solid-state nuclear magnetic resonance spectroscopy and relaxometry. *J. Pharm. Sci.* **2014**, *103*, 2635–2662. [CrossRef] [PubMed]
55. Thrane, L.W.; Berglund, E.A.; Wilking, J.N.; Vodak, D.; Seymour, J.D. NMR relaxometry to characterize the drug structural phase in a porous construct. *Mol. Pharm.* **2018**, *15*, 2614–2620. [CrossRef]
56. Yang, F.; Su, Y.; Zhu, L.; Brown, C.D.; Rosen, L.A.; Rosenery, K.J. Rheological and solid-state NMR assessments of copovidone/clotrimazole model solid dispersions. *Int. J. Pharm.* **2016**, *500*, 20–31. [CrossRef] [PubMed]
57. Carpentier, L.; Decressain, R.; Gusseme, A.; Neves, C.; Descamps, M. Molecular mobility in glass forming fananserine: A dielectric, NMR, and TMDSC investigation. *Pharm. Res.* **2006**, *23*, 798–805. [CrossRef] [PubMed]
58. Dudek, M.K.; Kamierski, S.; Potrzebowski, M.J. Fast and very fast MAS solid state NMR studies of pharmaceuticals. *Annu. Rep. NMR Spectro.* **2021**, *103*, 97–189.
59. Yuan, X.; Sperger, D.; Munson, E.J. Investigating miscibility and molecular mobility of nifedipine-PVP amorphous solid dispersions using solid-state NMR spectroscopy. *Mol. Pharm.* **2014**, *11*, 329–337. [CrossRef] [PubMed]
60. Sarpal, K.; Tower, C.W.; Munson, E.J. Investigation into intermolecular interactions and phase behavior of binary and ternary amorphous solid dispersions of ketoconazole. *Mol. Pharm.* **2020**, *17*, 787–801. [CrossRef]
61. Lu, X.; Huang, C.; Lowinger, M.B.; Yang, F.; Xu, W.; Brown, C.D.; Hesk, D.; Koynov, A.; Schenck, L.; Su, Y. Molecular Interactions in Posaconazole Amorphous Solid Dispersions from Two-Dimensional Solid-State NMR Spectroscopy. *Mol. Pharm.* **2019**, *16*, 2579–2589. [CrossRef] [PubMed]
62. Lu, X.; Li, M.; Huang, C.; Lowinger, M.B.; Xu, W.; Yu, L.; Byrn, S.R.; Templeton, A.C.; Su, Y. Atomic-Level Drug Substance and Polymer Interaction in Posaconazole Amorphous Solid Dispersion from Solid-State NMR. *Mol. Pharm.* **2020**, *17*, 2585–2598. [CrossRef] [PubMed]
63. Qian, F.; Huang, J.; Zhu, Q.; Haddadin, R.; Gawel, J.; Garmise, R.; Hussain, M. Is a distinctive single Tg a reliable indicator for the homogeneity of amorphous solid dispersion? *Int. J. Pharm.* **2010**, *395*, 232–235. [CrossRef]
64. Paudel, A.; Van Humbeeck, J.; Van den Mooter, G. Theoretical and experimental investigation on the solid solubility and miscibility of naproxen in poly(vinylpyrrolidone). *Mol. Pharm.* **2010**, *7*, 1133–1148. [CrossRef]
65. Litvinov, V.M.; Guns, S.; Adriaensens, P.; Scholtens, B.J.; Quaedflieg, M.P.; Carleer, R.; Van den Mooter, G. Solid state solubility of miconazole in poly[(ethylene glycol)-g-vinyl alcohol] using hot-melt extrusion. *Mol. Pharm.* **2012**, *9*, 2924–2932. [CrossRef]
66. Sarpal, K.; Delaney, S.; Zhang, G.G.; Munson, E.J. Phase Behavior of Amorphous Solid Dispersions of Felodipine: Homogeneity and Drug−Polymer Interactions. *Mol. Pharm.* **2019**, *16*, 4836–4851. [CrossRef]
67. Sarpal, K.; Munson, E.J. Amorphous Solid Dispersions of Felodipine and Nifedipine with Soluplus®: Drug-Polymer Miscibility and Intermolecular Interactions. *J. Pharm. Sci.* **2021**, *110*, 1457–1469. [CrossRef]
68. Duan, P.; Lamm, M.S.; Yang, F.; Xu, W.; Skomski, D.; Su, Y.; Schmidt-Rohr, K. Quantifying molecular mixing and heterogeneity in pharmaceutical dispersions at sub-100 nm resolution by spin diffusion NMR. *Mol. Pharm.* **2020**, *17*, 3567–3580. [CrossRef] [PubMed]
69. Ricarte, R.G.; Van Zee, N.J.; Li, Z.; Johnson, L.M.; Lodge, T.P.; Hillmyer, M.A. Recent advances in understanding the micro- and nanoscale phenomena of amorphous solid dispersions. *Mol. Pharm.* **2019**, *16*, 4089–4103. [CrossRef]
70. Knapik-Kowalczuk, J.; Rams-Baron, M.; Paluch, M. Current research trends in dielectric relaxation studies of amorphous pharmaceuticals: Physical stability, tautomerism, and the role of hydrogen bonding. *TrAC-Trend Anal. Chem.* **2021**, *134*, 116097. [CrossRef]
71. Kothari, K.; Ragoonanan, V.; Suryanarayanan, R. Influence of molecular mobility on the physical stability of amorphous pharmaceuticals in the supercooled and glassy States. *Mol. Pharm.* **2014**, *11*, 3048–3055. [CrossRef]

72. Bhardwaj, S.P.; Arora, K.K.; Kwong, E.; Templeton, A.; Clas, S.D.; Suryanarayanan, R. Mechanism of amorphous itraconazole stabilization in polymer solid dispersions: Role of molecular mobility. *Mol. Pharm.* **2014**, *11*, 4228–4237. [CrossRef] [PubMed]
73. Bhardwaj, S.P.; Suryanarayanan, R. Molecular mobility as an effective predictor of the physical stability of amorphous trehalose. *Mol. Pharm.* **2012**, *9*, 3209–3217. [CrossRef] [PubMed]
74. Dantuluri, A.K.; Amin, A.; Puri, V.; Bansal, A.K. Role of alpha-relaxation on crystallization of amorphous celecoxib above T(g) probed by dielectric spectroscopy. *Mol. Pharm.* **2011**, *8*, 814–822. [CrossRef]
75. Kothari, K.; Ragoonanan, V.; Suryanarayanan, R. The role of polymer concentration on the molecular mobility and physical stability of nifedipine solid dispersions. *Mol. Pharm.* **2015**, *12*, 1477–1484. [CrossRef]
76. Kothari, K.; Ragoonanan, V.; Suryanarayanan, R. The role of drug-polymer hydrogen bonding interactions on the molecular mobility and physical stability of nifedipine solid dispersions. *Mol. Pharm.* **2015**, *12*, 162–170. [CrossRef]
77. Mohapatra, S.; Samanta, S.; Kothari, K.; Mistry, P.; Suryanarayanan, R. Effect of polymer molecular weight on the crystallization behavior of indomethacin amorphous solid dispersions. *Cryst. Growth Des.* **2017**, *17*, 3142–3150. [CrossRef]
78. Mistry, P.; Suryanarayanan, R. Strength of drug-polymer interactions: Implications for crystallization in dispersions. *Cryst. Growth Des.* **2016**, *16*, 5141–5149. [CrossRef]
79. Mehta, M.; Kothari, K.; Ragoonanan, V.; Suryanarayanan, R. Effect of water on molecular mobility and physical stability of amorphous pharmaceuticals. *Mol. Pharm.* **2016**, *13*, 1339–1346. [CrossRef] [PubMed]
80. Mehta, M.; Suryanarayanan, R. Accelerated physical stability testing of amorphous dispersions. *Mol. Pharm.* **2016**, *13*, 2661–2666. [CrossRef] [PubMed]
81. Fung, M.H.; Suryanarayanan, R. Use of a plasticizer for physical atability prediction of amorphous solid dispersions. *Cryst. Growth Des.* **2017**, *17*, 4315–4325. [CrossRef]
82. Madejczyk, O.; Kaminska, E.; Tarnacka, M.; Dulski, M.; Jurkiewicz, K.; Kaminski, K.; Paluch, M. Studying the crystallization of various polymorphic forms of nifedipine from binary mixtures with the use of different experimental techniques. *Mol. Pharm.* **2017**, *14*, 2116–2125. [CrossRef] [PubMed]
83. Bhattacharya, S.; Suryanarayanan, R. Local mobility in amorphous pharmaceuticals-characterization and implications on stability. *J. Pharm. Sci.* **2009**, *98*, 2935–2953. [CrossRef]
84. Mehta, M.; Ragoonanan, V.; McKenna, G.B.; Suryanarayanan, R. Correlation between molecular mobility and physical stability in pharmaceutical glasses. *Mol. Pharm.* **2016**, *13*, 1267–1277. [CrossRef]
85. Knapik, J.; Wojnarowska, Z.; Grzybowska, K.; Jurkiewicz, K.; Tajber, L.; Paluch, M. Molecular dynamics and physical stability of coamorphous ezetimib and indapamide mixtures. *Mol. Pharm.* **2015**, *12*, 3610–3619. [CrossRef]
86. Knapik-Kowalczuk, J.; Wojnarowska, Z.; Rams-Baron, M.; Jurkiewicz, K.; Cielecka-Piontek, J.; Ngai, K.L.; Paluch, M. Atorvastatin as a promising crystallization inhibitor of amorphous probucol: Dielectric studies at ambient and elevated pressure. *Mol. Pharm.* **2017**, *14*, 2670–2680. [CrossRef] [PubMed]
87. Knapik-Kowalczuk, J.; Tu, W.; Chmiel, K.; Rams-Baron, M.; Paluch, M. Co-stabilization of amorphous pharmaceuticals-The case of nifedipine and nimodipine. *Mol. Pharm.* **2018**, *15*, 2455–2465. [CrossRef]
88. Fung, M.H.; Berzins, K.; Suryanarayanan, R. Physical stability and dissolution behavior of ketoconazole-organic acid coamorphous systems. *Mol. Pharm.* **2018**, *15*, 1862–1869. [CrossRef]
89. Fung, M.H.; DeVault, M.; Kuwata, K.T.; Suryanarayanan, R. Drug-excipient interactions: Effect on molecular mobility and physical stability of ketoconazole-organic acid coamorphous systems. *Mol. Pharm.* **2018**, *15*, 1052–1061. [CrossRef] [PubMed]
90. Knapik, J.; Wojnarowska, Z.; Grzybowska, K.; Jurkiewicz, K.; Stankiewicz, A.; Paluch, M. Stabilization of the amorphous ezetimibe drug by confining its dimension. *Mol. Pharm.* **2016**, *13*, 1308–1316. [CrossRef]
91. Zhang, C.; Sha, Y.; Zhang, Y.; Cai, T.; Li, L.; Zhou, D.; Wang, X.; Xue, G. Nanostructures and dynamics of isochorically confined amorphous drug mediated by cooling rate, interfacial, and intermolecular interactions. *J. Phys. Chem. B* **2017**, *121*, 10704–10716. [CrossRef]
92. Thakral, S.; Terban, M.W.; Thakral, N.K.; Suryanarayanan, R. Recent advances in the characterization of amorphous pharmaceuticals by X-ray diffractometry. *Adv. Drug Deliv. Rev.* **2016**, *100*, 183–193. [CrossRef]
93. Hedoux, A. Recent developments in the Raman and infrared investigations of amorphous pharmaceuticals and protein formulations: A review. *Adv. Drug Deliv. Rev.* **2016**, *100*, 133–146. [CrossRef]
94. Duggirala, N.K.; Li, J.; Kumar, N.S.K.; Gopinath, T.; Suryanarayanan, R. A supramolecular synthon approach to design amorphous solid dispersions with exceptional physical stability. *Chem. Commun.* **2019**, *55*, 5551–5554. [CrossRef] [PubMed]
95. Sahoo, A.; Kumar, N.S.K.; Suryanarayanan, R. Crosslinking: An avenue to develop stable amorphous solid dispersion with high drug loading and tailored physical stability. *J. Control Release* **2019**, *311–312*, 212–224. [CrossRef] [PubMed]
96. Li, N.; Taylor, L.S. Nanoscale infrared, thermal, and mechanical characterization of telaprevir-polymer miscibility in amorphous solid dispersions prepared by solvent evaporation. *Mol. Pharm.* **2016**, *13*, 1123–1136. [CrossRef] [PubMed]
97. Purohit, H.S.; Taylor, L.S. Miscibility of itraconazole-hydroxypropyl methylcellulose blends: Insights with high resolution analytical methodologies. *Mol. Pharm.* **2015**, *12*, 4542–4553. [CrossRef]
98. Sibik, J.; Zeitler, J.A. Direct measurement of molecular mobility and crystallisation of amorphous pharmaceuticals using terahertz spectroscopy. *Adv. Drug Deliv. Rev.* **2016**, *100*, 147–157. [CrossRef]

99. Chen, Z.; Yang, K.; Huang, C.; Zhu, A.; Yu, L.; Qian, F. Surface enrichment and depletion of the active ingredient in spray dried amorphous solid dispersions. *Pharm. Res.* **2018**, *35*, 38. [CrossRef]
100. Rautaniemi, K.; Vuorimaa-Laukkanen, E.; Strachan, C.J.; Laaksonen, T. Crystallization kinetics of an amorphous pharmaceutical compound using fluorescence lifetime imaging microscopy. *Mol. Pharm.* **2018**, *15*, 1964–1971. [CrossRef]
101. Bhujbal, S.V.; Zemlyanov, D.; Cavallaro, A.A.; Mangal, S.; Taylor, L.S.; Zhou, Q.T. Qualitative and quantitative characterization of composition heterogeneity on the surface of spray dried amorphous solid dispersion particles by an advanced surface analysis platform with high surface-sensitivity and superior spatial resolution. *Mol. Pharm.* **2018**, *15*, 2045–2053. [CrossRef] [PubMed]

Article

Entrapped Transient Chloroform Solvates of Bilastine

Cristina Puigjaner [1,*], Anna Portell [1], Arturo Blasco [2], Mercè Font-Bardia [1] and Oriol Vallcorba [3]

1. Centres Científics i Tecnològics, Universitat de Barcelona, Lluís Solé i Sabarís 1-3, 08028 Barcelona, Spain; portell@ccit.ub.edu (A.P.); mercef@ccit.ub.edu (M.F.-B.)
2. Institut Químic de Sarrià, Universitat Ramon Llull, Via Augusta 390, 08017 Barcelona, Spain; arturoblascos@iqs.edu
3. ALBA Synchrotron Light Source, Carrer de la Llum 2-26, Cerdanyola del Vallès, 08290 Barcelona, Spain; ovallcorba@cells.es
* Correspondence: cris@ccit.ub.edu; Tel.: +34-93-4021692

Abstract: The knowledge about the solid forms landscape of Bilastine (BL) has been extended. The crystal structures of two anhydrous forms have been determined, and the relative thermodynamic stability among the three known anhydrous polymorphs has been established. Moreover, three chloroform solvates with variable stoichiometry have been identified and characterized, showing that $S_{3CHCl3-H2O}$ and S_{CHCl3} can be classified as transient solvates which transform into the new chloroform solvate $S_{CHCl3-H2O}$ when removed from the mother liquor. The determination of their crystal structures from combined single crystal/synchrotron X-ray powder diffraction data has allowed the complete characterization of these solvates, being two of them heterosolvates ($S_{3CHCl3-H2O}$ and $S_{CHCl3-H2O}$) and S_{CHCl3} a monosolvate. Moreover, the temperature dependent stability and interrelation pathways among the chloroform solvates and the anhydrous forms of BL have been studied.

Keywords: solvate; polymorph; bilastine; crystal structure determination; single crystal X-ray diffraction; structure determination from powder diffraction

Citation: Puigjaner, C.; Portell, A.; Blasco, A.; Font-Bardia, M.; Vallcorba, O. Entrapped Transient Chloroform Solvates of Bilastine. *Crystals* **2021**, *11*, 342. https://doi.org/10.3390/cryst11040342

Academic Editor: Venu Vangala

Received: 14 March 2021
Accepted: 24 March 2021
Published: 28 March 2021

Publisher's Note: MDPI stays neutral with regard to jurisdictional claims in published maps and institutional affiliations.

Copyright: © 2021 by the authors. Licensee MDPI, Basel, Switzerland. This article is an open access article distributed under the terms and conditions of the Creative Commons Attribution (CC BY) license (https://creativecommons.org/licenses/by/4.0/).

1. Introduction

A search for the different solid forms of an active pharmaceutical ingredient (API) is a crucial part of the drug development process. Solvate formation has many implications in the pharmaceutical industry, as it affects the physicochemical properties of materials, such as their melting point, density and dissolution rate, which in turn can influence their manufacturability and pharmacokinetic properties, without changing the pharmacology of the API through modification of covalent bonds [1–4]. The unexpected formation of undesired solvates can thus lead to unpredictable behavior of the drug and could prove costly. Organic solvents are constantly present in the pharmaceutical production processes, so many aspects of them have to be extremely controlled [5].

The rationalization of the solvate formation is one of the important topics in the current crystal engineering [6], and the discovery of solvates is important in several aspects: (1) formation of solvates can limit the selection of solvents for crystallization of the desired crystal form; (2) solvates can be used as intermediates for producing the necessary polymorphs, as specific polymorphs can sometimes be obtained only via desolvation of particular solvates; (3) solvates can serve to control the particle size distribution in the product in cases in which the nonsolvated forms are difficult to crystallize [6]; and (4) particularly stable solvates, typically but not exclusively hydrates (like diosgenin hydrate [7] for example), can be used as the marketed form [8]. Although the utilization of the most physically stable crystal form is typically desired as any change in the solid form may affect the bioavailability associated with the drug product, metastable modifications can be preferred when an improvement in the in vitro dissolution kinetics is achieved. Up to the present, there are some few solvates on the market such as trametinib dimethyl sulfoxide, dapagliflozin propanediol monohydrate, cabazitaxel, darunavir ethanolate, warfarin sodium, indinavir sulfate ethanolate and atorvastatin calcium [9]. Limitations for the use of solvates in pharmaceutical industry are given by the toxicity of solvents they contain, and also, they may additionally accelerate decomposition of the final product. Permitted solvents and the limits of their content are provided by the regulatory authorities in the pharmacopeias [10].

The general prediction of solvate formation, similar to the prediction of other solid forms, is still largely an unresolved problem. Currently, in order to avoid unexpected structural transformations such as hydrate and solvate forms, in the pharmaceutical industry, high-throughput crystallization experiments are conducted to obtain all possible solid forms of a drug [11].

Two main structural driving forces responsible for incorporation of solvent molecules in the structure have been identified, on one hand, the ability of solvents to compensate unsatisfied potential intermolecular interactions between the molecules and, on the other hand, the ability to decrease the void space and/or lead to more efficient packing. Both the formation of an extensive hydrogen bond network established by the solvent molecules as well as an increase of the packing efficiency have been shown to be the main contributing factors for the solvate formation of pharmaceutical molecules. Most of the solvates, however, include contributions from both of these driving forces, and the solvate formation thus is due to a lowering of the crystal free energy [8].

Bilastine, 2-[4-[2-[4-[1-(2-ethoxyethyl)-1H-benzimidazol-2-yl]-1-piperidinyl]ethyl] phenyl]-2-methylpropionic acid (Figure 1), is a well-tolerated, second generation antihistamine drug approved for the symptomatic treatment of allergic rhinoconjunctivitis and chronic urticaria [12]. It exerts its effect as a selective histamine H_1 receptor antagonist and has an effectiveness similar to cetirizine, fexofenadine and desloratadine [13]. It was developed in Spain by FAES Farma, and it has been commercially available internationally since March 2011.

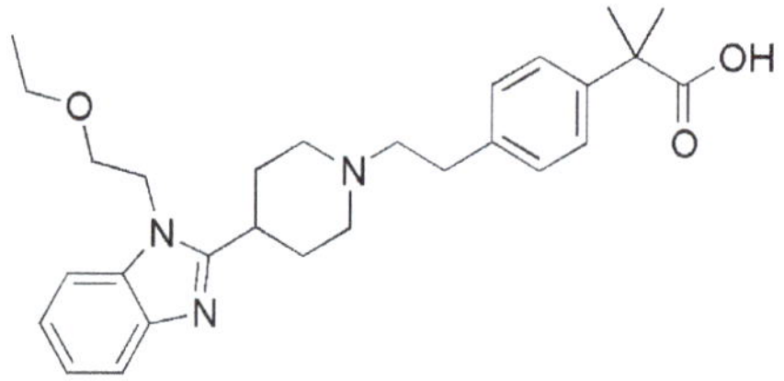

Figure 1. Chemical structure of Bilastine.

BL, its preparation and uses as H1 receptor antagonist were first described in the European patent EP0818454B1 [14]. Later, the patent WO03089425 reported three crystalline forms of BL: 1, 2 and 3, characterized by IR, and crystallographic parameters were provided only for form 1 [15]. It was said that forms 2 and 3 of BL easily converted into form 1. In the present study we have extended the knowledge about the solid forms landscape of BL by performing a polymorph screening starting from anhydrous forms I and III, whose crystal structures have been determined from single crystal and synchrotron powder X-ray diffraction, respectively. The thermodynamic relationship among the anhydrous forms has been established. In addition, during the screening three chloroform solvates (two heterosolvates ($S_{3CHCl3-H2O}$ and $S_{CHCl3-H2O}$) and one monosolvate S_{CHCl3}) have been obtained, being two of them transient solvates which transform into solvate $S_{CHCl3-H2O}$ immediately when exposed to ambient conditions. The crystal structures of the three solvates with different stoichiometries have been determined and will be discussed and compared with the anhydrous forms. Moreover, the different forms have been further characterized by differential scanning calorimetry (DSC), thermogravimetric analysis (TGA) and variable temperature powder X-ray diffraction (VT-PXRD). Finally, the phase transformations pathways among the different forms have been defined.

2. Materials and Methods

2.1. Preparation of Bilastine Crystal Forms

The organic solvents used were all of analytical quality. The screen included evaporations, cooling crystallizations at different rates, antisolvent precipitations, antisolvent diffusions and slurries (see details in the Supplementary Materials). The experimental solid form screening resulted in six solid forms (three anhydrates and three chloroform solvates) as confirmed by PXRD and thermal analysis.

2.1.1. Form I

BL (40 mg) (0.086 mmol) was dissolved in DCM (0.7 mL) at 60 °C, and the solution was slowly cooled down to 25 °C. After 18 days, single crystals were obtained, filtered, dried under vacuum and analyzed by SCXRD.

2.1.2. Form II

Form III of BL (20 mg) (0.043 mmol) was placed in a 70 µL alumina crucible and heated under nitrogen atmosphere inside a TGA equipment from 30 °C to 200 °C at a rate of 10 °C/min, maintained at 200 °C for 5 min and cooled down to 30 °C.

2.1.3. Form III

BL (100 mg) (0.216 mmol) was dissolved in $CHCl_3$ (0.6 mL) at 60 °C, and the solution was cooled down to 25 °C outside the heating block. The solid precipitated after one night and it was filtered and dried under vacuum.

2.1.4. $S_{3CHCl3-H2O}$

$CHCl_3$ (2.3 mL) was added to 330 mg of BL at 60 °C. The solution was cooled down overnight and left in the fridge (4–5 °C) for four days until single crystal growing. This solvate was stable in the mother liquor. The single crystal was analyzed by SCXRD at low temperature to prevent its transformation into form $S_{CHCl3-H2O}$.

2.1.5. $S_{CHCl3-H2O}$

$CHCl_3$ (6.5 mL) was added to 1 g of BL, and the resulting suspension was stirred overnight at r.t. The solid was filtered and dried under vacuum.

2.1.6. S_{CHCl3}

BL (45 mg) was dissolved in the minimum quantity of anhydrous $CHCl_3$ (0.35 mL) at 90 °C. The solution was cooled down inside the fridge (4–5 °C) until single crystal growing. This solvate was stable in the mother liquor. The single crystal was analyzed by SCXRD at low temperature to prevent its transformation into form $S_{CHCl3-H2O}$.

2.2. Methods

2.2.1. Powder X-ray Diffraction (PXRD)

Powder X-ray diffraction patterns were obtained on a PANalytical X'Pert PRO MPD diffractometer (Malvern Panalytical, Almelo, Netherlands) in transmission configuration using Cu Kα1+2 radiation (λ = 1.5418 Å) with a focalizing elliptic mirror and a PIXcel detector working at a maximum detector's active length of 3.347°. Capillary geometry has been used with samples placed in glass capillaries (Lindemman) of 0.5 mm of diameter measuring from 2 to 60° in 2θ, with a step size of 0.026° and a total measuring time of 30 min. Flat geometry has been used for routine samples sandwiched between low absorbing films (polyester of 3.6 microns of thickness) measuring 2theta/theta scans from 2 to 40° in 2θ with a step size of 0.026° and a measuring time of 80 seconds per step.

Powder X-ray diffraction patterns of forms III and $S_{CHCl3-H2O}$ were obtained using synchrotron radiation at ALBA's beam line BL04-MSPD using Mythen detector [16]. The wavelength, 0.63493 Å for form III and 0.61927 Å for $S_{CHCl3-H2O}$, was selected with a double-crystal Si (111) monochromator and determined from a Si640d NIST standard (a = 5.43123 Å) measurement. The diffractometer is equipped with a so-called MYTHEN detector system especially suited for time-resolved experiments. The capillary of 0.7 mm containing the sample was rotated during data collection to improve diffracting particle statistics. The data acquisition time was 10 min per pattern, and the final treated data are the addition of ten acquisitions to attain very good signal-to-noise ratio over the angular range 0.5–43.6° (2θ) at 300K. Both powder diffraction patterns were indexed using DICVOL06 [17], and the obtained cell parameters were refined via pattern matching with Dajust software [18]. The crystal structures were solved with the direct-space strategy TALP [19] introducing the bond distances and angles from the single crystal structure of form I as restraints. Crystal structure solution was followed by the location of the solvent molecules in the difference electron density map and a restrained Rietveld refinement with RIBOLS to obtain the final crystal structures. H-atoms were placed to calculated positions after the final refinement.

Variable temperature experiments of $S_{3CHCl3-H2O}$ sample were performed at the ALBA Synchrotron using a Cyberstar hot gas blower with an Eurotherm temperature controller (Eurotherm, Worthing, UK) and continuously collecting diffraction data during the heating and cooling ramps (30° to 230° and 230° to 30° at 2°/min, 18 sec/pattern). A longer measurement (180s) was performed at room temperature.

2.2.2. Single Crystal X-ray Diffraction (SCXRD)

The single crystal structures were solved on a D8 (Bruker, Karlsrühe, Germany) Venture system equipped with a multilayer monochromator, and a Mo microfocus (λ = 0.71073 Å) has been used too. Frames were integrated with the Bruker SAINT software package using a SAINT algorithm. Data were corrected for absorption effects using the multi-scan method (SADABS). The structures were solved and refined using the Bruker SHELXTL Software Package [20], a computer program for automatic solution of crystal structure, and refined by full-matrix least-squares method with ShelXle Version 4.8.0, a Qt graphical user interface for SHELXL computer program [21].

2.2.3. Differential Scanning Calorimetry (DSC)

Differential scanning calorimetry was carried out by means of a Mettler-Toledo DSC-822e calorimeter (Mettler-ToledoAG, Schwerzenbach, Switzerland). Experimental conditions: Aluminum crucibles of 40 µL volume, atmosphere of dry nitrogen with 50 mL/min flow rate, heating rate of 10 °C/min. The calorimeter was calibrated with indium of 99.99% purity.

2.2.4. Thermogravimetric Analysis (TGA)

Thermogravimetric analyses were performed to detect the presence of hydrates/solvates on a Mettler-Toledo TGA-851e thermobalance (Mettler-ToledoAG, Schwerzenbach, Switzerland). Experimental conditions: Alumina crucibles of 70 µL volume, atmosphere of dry nitrogen with 50 mL/min flow rate, heating rate of 10 °C/min.

3. Results and Discussion

3.1. Anhydrous Forms

Three anhydrous forms of BL have been identified. According to their melting points (Figure 2), forms I, II and III correspond to polymorphs 1, 2 and 3 previously reported in patent WO03089425. The DSC of the lowest melting form III shows its melting and simultaneous crystallization and melting of form I followed by the crystallization and melting of form II, the highest melting form.

High quality single crystals suitable for crystal structure determination have been obtained only for Form I, while the crystal structure of form III was determined from PXRD data obtained in Synchrotron. Regarding form II, only its unit cell and space group could be determined. The X-ray powder diffractograms of the three anhydrous forms are shown in Figure 3.

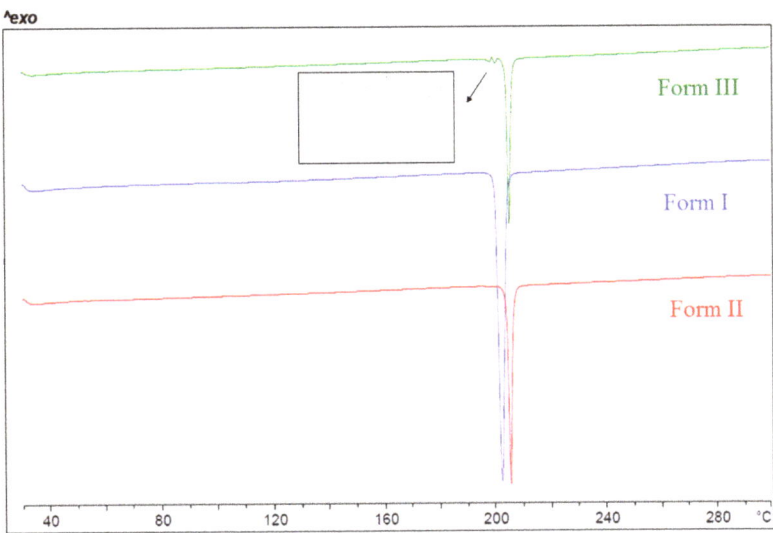

Figure 2. DSC analysis of the different anhydrous forms of BL, showing the melting of form I at 200 °C, form II at 202 °C and form III at 197 °C followed by simultaneous crystallization and melting of form I and subsequent crystallization and melting of form II.

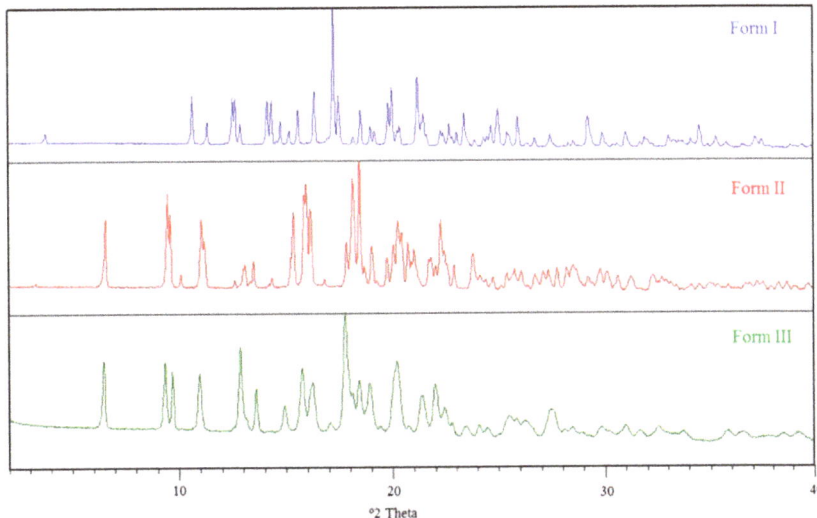

Figure 3. PXRD patterns of the different anhydrous forms of BL.

3.2. Relative Thermodynamic Stability among the Anhydrous Solid Forms

Form I was known to be the most stable form at r.t., and slurries of mixtures of I + II and I + III resulted in complete transformation into form I. The main question to solve is whether two polymorphs are monotropically (one form is more stable than the other at any temperature) or enantiotropically (a transition temperature exists, below and above which the stability order is reversed) related, and for an enantiotropic system, where transition temperature lies.

Table 1 summarizes the physicochemical data of the three modifications of BL obtained from the thermal analysis experiments. According to the Heat of Fusion Rule of Burger and Ramberger [22], forms I and II are enantiotropically related due to the higher enthalpy of fusion and lower melting point of form I. An estimation of the transition temperature between an enantiotropic pair of polymorphs can be done by using the treatment of Yu [23]. This method uses the temperatures and enthalpies of fusion to calculate the Gibbs free energy difference at the melting temperature of the lower melting form and to extrapolate to other temperatures.

$$T_{trs} = \frac{\Delta H_{fus,2} - \Delta H_{fus,1} + k\Delta H_{fus,1}\left(T_{fus,1} - T_{fus,2}\right)}{\frac{\Delta H_{fus,2}}{T_{fus,2}} - \frac{\Delta H_{fus,1}}{T_{fus,1}} + k\Delta H_{fus,1} \ln\left(\frac{T_{fus,1}}{T_{fus,2}}\right)} \quad (1)$$

Table 1. Physicochemical data of anhydrous forms of BL.

	T_{fus} (°C)	ΔH_{fus} (J/g)
Form I	200	119.9
Form II	202	118.2
Form III	197	-

A value of 0.003 was used for the factor k, which was empirically determined and allows a good approximation of the heat capacity differences in the majority of cases [23]. Using this equation, we calculated a value of 137 °C for the I/II pair.

It was not possible to obtain the value of the enthalpy of fusion of form III as this form transformed while heating in the DSC analysis. Regarding the relative thermodynamic stability between the pair I/III, form I is more stable than III at r.t., being this last one the lower melting form. Therefore, forms I and III must be monotropically related at least from r.t., being form I the most stable one.

As for the pair II/III, without knowing the enthalpy of fusion of form III, it was not possible to use the Heat of Fusion Rule of Burger and Ramberger; therefore, we applied the solvent mediated transformation method [24]. This method is based on the relationship between solubility and stability of crystal forms, i.e., the less stable form will also be the most soluble at given conditions of temperature and pressure. If crystals of both forms are mixed with a saturated solution of the product, the most stable form will grow at the expense of the less stable one. Thus, a mixture of the two modifications was stirred in ethanol at r.t., and after one hour, pure form I was recovered. Hence, a reduction of time was required and a suspension of II and III was stirred for 15 min. Comparing the PXRD between the initial and the final mixture, a decrease in the intensity of X-ray diffraction peaks attributed to form II (compared to form III) was observed, while form I did not appear. This final mixture was suspended again in ethanol for 10 min, and form II decreased again; however, form I began to appear in the mixture, so the transformation of II into III no longer could be studied. We concluded that form III was more stable than form II at r.t., and as III is the lower melting form, an enantiotropy relation could be inferred for the II/III pair. Without having data of the melting enthalpy of form III, it was not possible to calculate the transition temperature between II and III.

Energy diagrams such as those proposed by Burger and Ramberger [22] give considerable insight into polymorphic systems. Based on physicochemical and solvent mediated transformations data, a semi-schematic energy/temperature diagram was constructed in order to display the thermodynamic relationship of the anhydrous polymorphs at different temperatures (Figure 4).

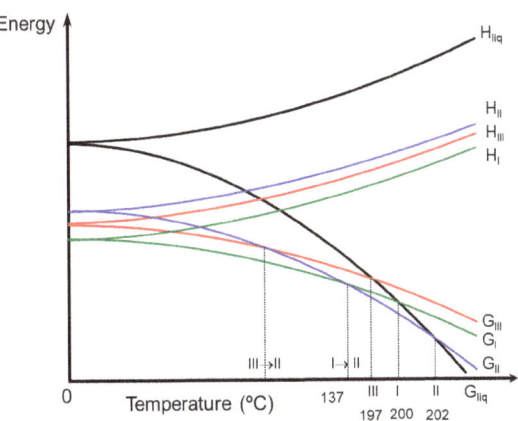

Figure 4. Semi-schematic energy/temperature diagram of anhydrous forms of BL.

3.3. Chloroform Solvates

During the preparation of the anhydrates, new forms were detected when chloroform was used. Therefore, a solid from screening starting from anhydrous form I and form III has been performed in chloroform as solvent. Different thermodynamic and kinetic conditions have been applied, and several antisolvents have been used. Only one chloroform solvate was identified by PXRD, and it has been fully characterized ($S_{CHCl3-H2O}$). Its DSC analysis (Figure 5) shows a first wide endothermic phenomenon overlapped with an endothermic phenomenon at 77 °C, which could be attributed to the evaporation of the solvent and the melting of the solvate taking place simultaneously. Thermal desolvation is a complex process in which the melting, release of solvent vapor and crystallization of a new phase may occur simultaneously. Next, a low intensity phenomenon is observed at 198 °C which could be attributed to the melting of form III, followed by the melting of form II at 203 °C. Its TGA analysis shows a weight loss of 19.4% from 32 to 100 °C which could be attributed to 1 chloroform molecule and 1 H_2O molecule (theoretical weight loss of 22.9%).

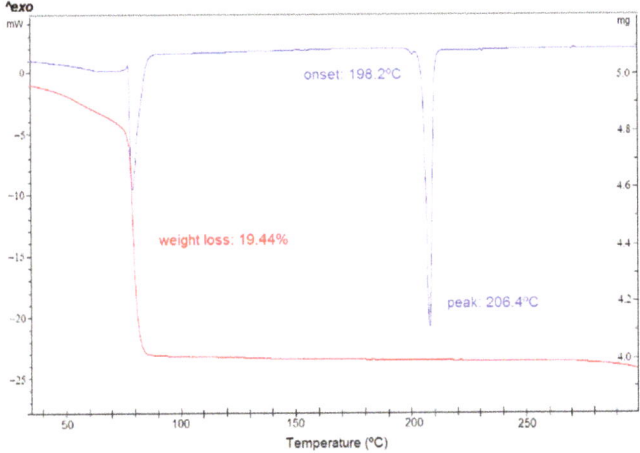

Figure 5. DSC (blue) and TGA (red) of form $S_{CHCl3-H2O}$.

Temperature variable powder X-ray diffraction of form $S_{CHCl3-H2O}$ has been performed in synchrotron Alba. In Figure 6, we can appreciate the diffractogram of $S_{CHCl3-H2O}$ which is stable until 74 °C and from this temperature, some new peaks corresponding to form

III begin to appear, while characteristic peaks of $S_{CHCl3-H2O}$ disappear. Form $S_{CHCl3-H2O}$ progressively transforms into form III while increasing temperature until 85 °C.

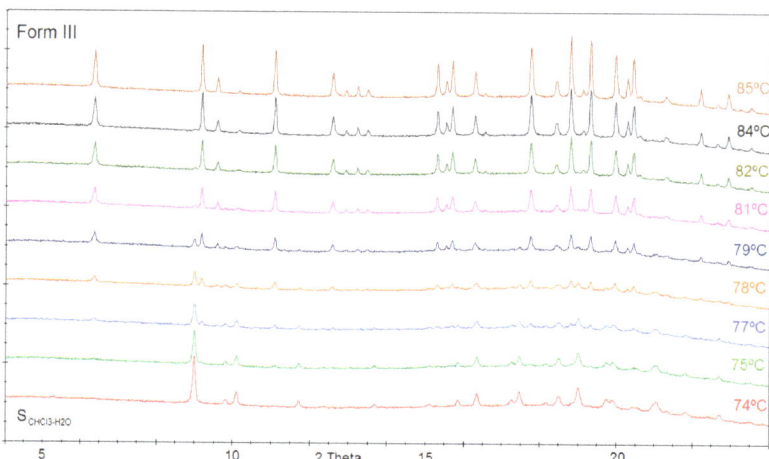

Figure 6. Temperature variable PXRD of $S_{CHCl3-H2O}$ showing its transformation into form III.

In addition, different experiments in chloroform (vapor diffusion, slow cool crystallizations) were performed in order to get a single crystal of $S_{CHCl3-H2O}$. Curiously, we always obtained under different conditions good quality crystals of a new solid form, as the PXRD calculated from its crystal structure solved did not match with any of the previously known forms. As shown by its crystal structure determination, this form consisted of one molecule of BL, one molecule of water and three chloroform molecules, so it was identified as the heterosolvate or mixed solvate $S_{3CHCl3-H2O}$. Its structure was solved at low temperature, and when leaving the crystal to reach r.t., the crystal collapsed, and it was impossible to measure it at ambient conditions. The presence of opaque crystals indicated that the product had transformed opaque due to pseudomorphosis. Therefore, the single crystals (which darkened immediately at r.t.) were analyzed by PXRD at r.t., and their diffractogram matched with the one of form $S_{CHCl3-H2O}$. Therefore, the new form obtained at low temperature in SCXRD is a precursor of $S_{CHCl3-H2O}$ by losing 2 chloroform molecules from its crystal structure. Indeed, form $S_{3CHCl3-H2O}$ is the one which is always obtained in chloroform solutions, and it immediately transforms into $S_{CHCl3-H2O}$ when removing it from the mother liquor, preventing its detailed characterization by other methods. The instability of this new identified solvate of BL means that we cannot exclude the discovery of other labile solvate forms. Highly unstable solvates are more common among small organic molecules than is generally believed [25].

The discovery of those mixed chloroform-water solvates led us to conduct a screening in anhydrous chloroform too. A new chloroform solvate S_{CHCl3} was discovered; however, it was only obtained as single crystal. Again, this solvate darkened when removing it from the mother liquor at r.t., and it was only possible to solve its structure at low temperature. The analysis of the corresponding PXRD at r.t. resulted in $S_{CHCl3-H2O}$, revealing that S_{CHCl3} transformed also into $S_{CHCl3-H2O}$, but in this case by incorporating one water molecule in its crystal structure. Figure 7 shows the PXRD diffractograms of the three chloroform solvates of BL.

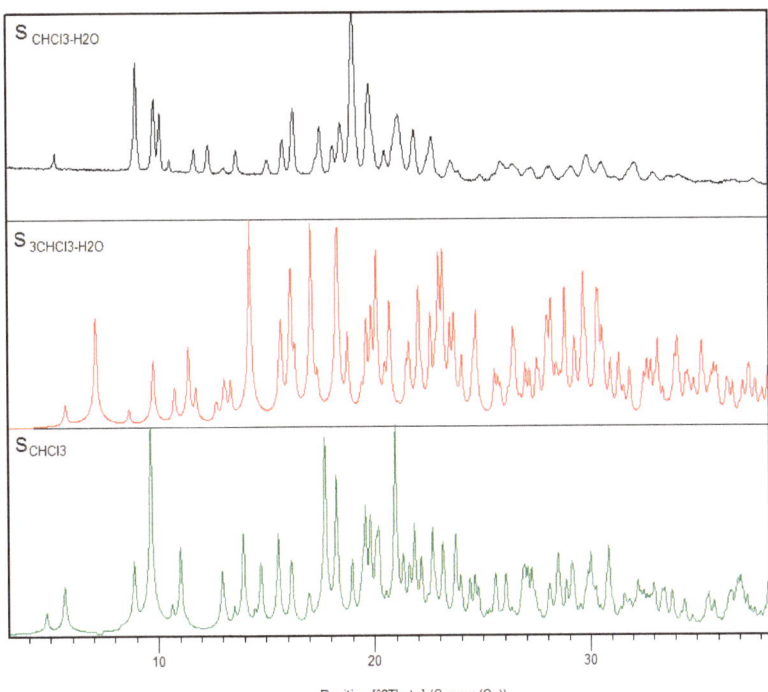

Figure 7. PXRD comparison among the different chloroform solvates.

Thus, we can conclude that both chloroform solvates S$_{3CHCl3·H2O}$ and S$_{CHCl3}$ are only stable at low temperature or inside chloroform solutions (anhydrous in the case of S$_{CHCl3}$) and that they transform immediately into S$_{CHCl3·H2O}$ at r.t. Both forms can be best described as transient or elusive phases as they become unstable after removal from the mother liquor. There are other examples in the literature of these labile solvates, and their existence needs to be known and considered [26]. Solvates that desolvate readily as soon as they are removed from the mother liquor can easily be overlooked in solid form screenings and wrong conclusions concerning the solvent effects on the nucleation of individual phases may result [27].

3.4. Stability and Interrelation Pathways among the Solvates and Anhydrous Forms

The stability of S$_{CHCl3·H2O}$ in front of temperature was studied when heating it in a DSC experiment, and it was observed that from 50 °C this solvate was no longer stable and began to transform into form III (Figure 5).

The stability of S$_{CHCl3·H2O}$ in front of relative humidity has been studied at r.t. by placing this sample inside desiccators containing different saturated salt solutions which create different RH conditions. From 0% to 43% RH, S$_{CHCl3·H2O}$ was stable at least during one month. At 57–75% RH, this form began to transform into form III after one month.

In addition, the stabilities of forms I, III and S$_{CHCl3·H2O}$ were studied under different chloroform atmospheres. While the relative thermodynamic stability of polymorphs depends only on the temperature at constant pressure, the stability relationship between a solvate and a non-solvated form, or two solvates, depends not only on the temperature but also on the activity of the solvent at constant pressure. To obtain different relative solvent vapor pressures, chloroform solutions in DMF were prepared according to the Raoult's law [28]:

$$x = \frac{p}{p_0} = \frac{n}{n_0 + n} \tag{2}$$

where x is the solvent mole fraction in solution; p_0 is the vapor pressure of pure solvent; p is the partial solvent vapor pressure over the solution; n represents the amount of used solvent in the solution; n_0 is the moles of DMF in the solution. The samples were placed in desiccators with the prepared solutions of different partial pressures of chloroform. At 0–40% CHCl$_3$, forms I and III remain invariable while $S_{CHCl3-H2O}$ transformed into II + traces of I at 0% and into I at 20–40% CHCl$_3$. While increasing chloroform proportion to 60–80%, form I did not change; form III transformed into a mixture of III+I, and $S_{CHCl3-H2O}$ remained invariable. Under 100% chloroform atmosphere, both anhydrous forms transformed into $S_{CHCl3-H2O}$.

Curiously, at r.t. under 0% RH, $S_{CHCl3-H2O}$ does not desolvate, while under 100% DMF atmosphere, mixture of II+traces of I is obtained.

The information obtained during the solid form screening and the study of the temperature- and moisture-dependent stability and interrelation pathways among the solid forms of BL has been summarized in Figure 8.

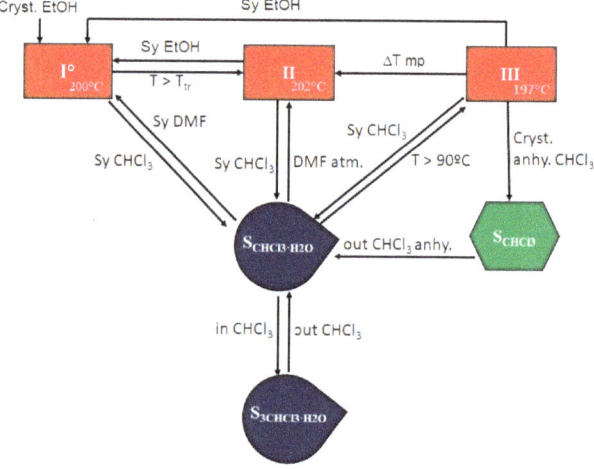

Figure 8. Pathways for phase transformations of BL solid forms as a result of heating or exposure to different solvents.

Form I is the most stable anhydrous form at r.t. which can be obtained by slurry in ethanol from the other anhydrous forms. Form II can be obtained from I and from III by increasing temperature, as it is enantiotropically related with both forms. A slurry in CHCl$_3$ at r.t. produces the heterosolvate $S_{3CHCl3-H2O}$ which is transformed into $S_{CHCl3-H2O}$ when taken out of the solution. Heating $S_{CHCl3-H2O}$ produces its desolvation and transformation into form III. On the other hand, the slurry of $S_{CHCl3-H2O}$ in DMF leads to form I while maintaining $S_{CHCl3-H2O}$ under DMF atmosphere allows the isolation of form II in short time (form I appears later). Finally, the use of anhydrous chloroform leads to chloroform solvate S_{CHCl3} which transforms also to $S_{CHCl3-H2O}$ when taken out of solution.

3.5. Crystal Structures Analysis

3.5.1. Bilastine Anhydrates

Three anhydrous polymorphs of BL were obtained. Form I was solved by single crystal XRD. This anhydrate crystallizes in the monoclinic space group $P2_1/c$ with Z = 4. This structure is zwitterionic as the piperidinyl N atom is protonated, revealing a proton transfer from the carboxylic group. For an amphoteric molecule with measured pK_a values of 4.06 (carboxylic acid) and 9.43 (piperidine-N), application of the pK_a rule of three would suggest that proton transfer from the carboxylic acid to the piperidine-N is essentially complete rendering BL zwitterionic. BL molecules are linked by hydrogen bonds (distance: 1.567 Å) involving the carboxylate oxygen of one BL molecule and the piperidinium nitrogen of another BL in an alternated mode, forming corrugated layer structures (Figure 9). π-π stackings are also observed between the benzene rings of the benzimidazoles (Figure 10). The observed lateral offset stack is 2.61 Å [29], the centroid to centroid distance, d(π···π) is 4.25 Å and the interplanar distance is 3.35 Å.

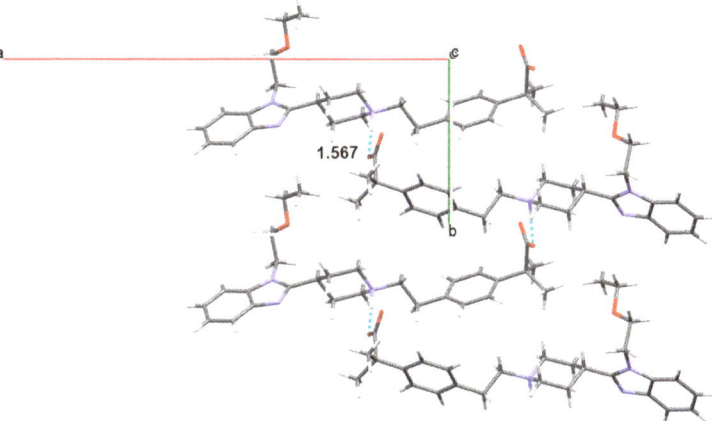

Figure 9. Carboxylate/piperidinium H-bond interactions (NH···O distance in Å) between BL molecules of Form I.

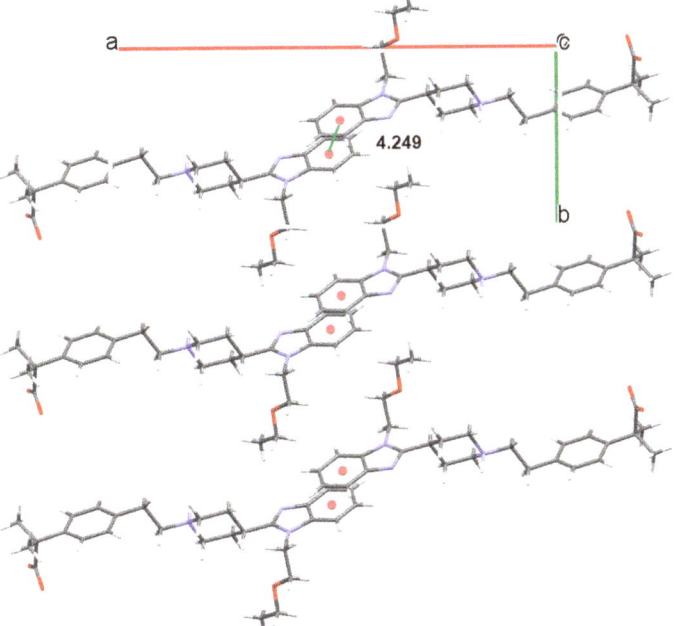

Figure 10. π-π stacking (centroid–centroid distance in Å) observed in Form I of BL.

Attempts to grow quality crystals of forms II and III were unsuccessful. Crystals of form III were grown and observed under the microscope. The extinction of a crystal is observed in Figure 11; however, it was impossible to cut the extreme of this crystal in order to determine its structure by SCXRD.

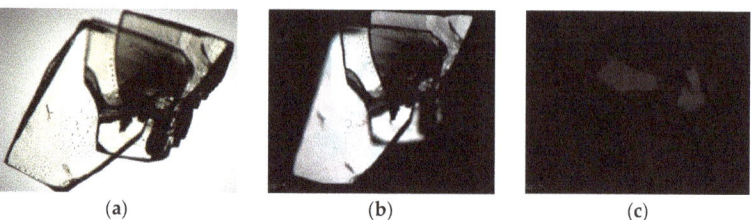

Figure 11. Photomicrographs of form III while looking under (**a**) unpolarized light and (**b**) polarized light on a circular microscope stage (**c**) Extinction of the crystal is observed at the optimum angle to display maximum birefringence.

Usually form III crystallized as macles, as it is shown in Figure 12.

Figure 12. Photomicrographs of macles of form III observed under (**a**) unpolarized light and (**b**) polarized light microscope.

Thus, the resolution of the crystal structure of form III was achieved by using the direct space methodology using synchrotron X-ray powder diffraction data obtained in the high-resolution powder diffraction end station of the MSPD beam line in Alba. The powder diffraction pattern of form III was indexed to a monoclinic cell of about 2682 Å^3, and the space group perfectly determined to be $P2_1/c$ from the systematic absences. The crystal structure was solved with the direct-space strategy TALP, and the refinement was performed by the Rietveld method. Figure 13 depicts the final Rietveld plot.

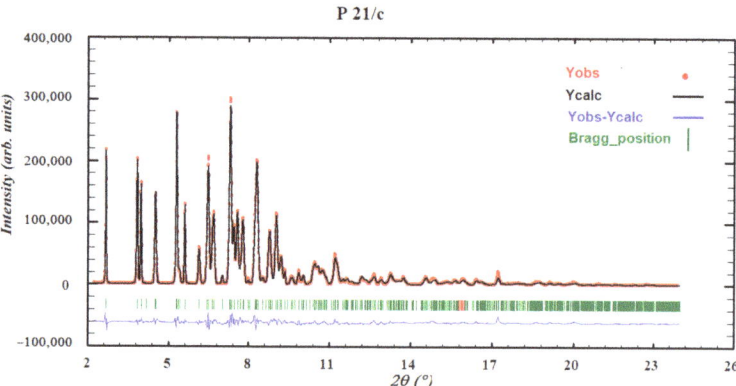

Figure 13. Final Rietveld plot for the crystal structure refinement of form III. Agreement factor: Rwp = 2.4%. The plot shows the experimental powder XRD profile (red+marks), the calculated powder XRD profile (black solid line), and the difference profile (blue, lower line). Tick marks indicate peak positions.

Form III crystallizes in the same space group $P2_1/c$ with four molecules in the unit cell. BL molecules in Form III are also linked by hydrogen bonds (distance: 1.518 Å) involving the carboxylate oxygen and the piperidinium nitrogen but not in the alternated mode shown in Form I. In Form III, all BL molecules are orientated in the same way. Therefore, a remarkable difference between both polymorphs is the different orientation of the self-assembled BL chains. While in Form I BL molecules are interdigitated, in Form III, they are organized as cascade layers (Figure 14). Moreover, in Form I benzimidazoles interact via π-π stacking while in Form III, the same benzimidazoles interact via CH-π with one hydrogen of the piperidinyl group (distance to the centroid: 2.93 Å, distance to the plane: 2.64 Å and H: 1.27 Å) and with one hydrogen of the methyl group (distance to the centroid: 3.41 Å, distance to the plane: 2.45 Å and H: 2.37 Å) (Figure 15).

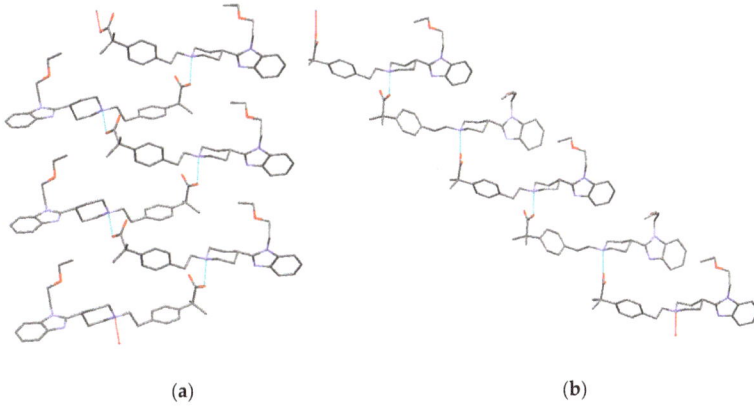

Figure 14. Comparison between the crystal structures of (**a**) form I (corrugated layer) and (**b**) form III (cascade layer) of BL.

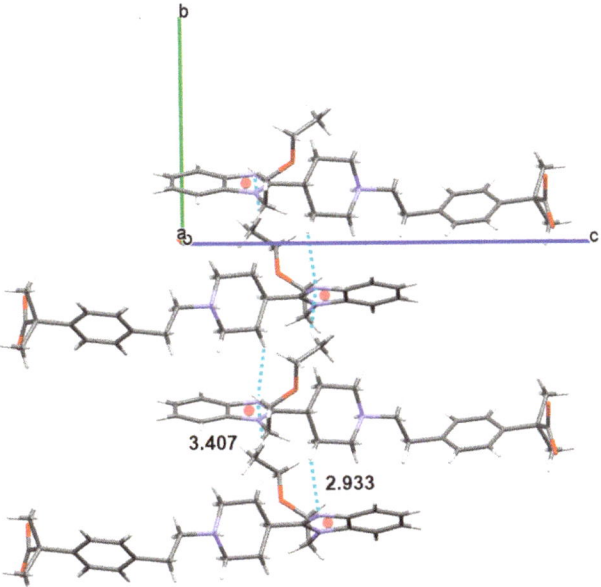

Figure 15. CH-π stacking (CH–centroid distances in Å) observed in Form III of BL.

The powder diffraction pattern of form II was indexed to a monoclinic cell (a = 28.3374 Å, b = 9.9384 Å, c = 19.7547 Å, β = 109.2696°) of about 5251.8 Å^3 and the space group perfectly determined to be $P2_1/c$ from the systematic absences (Figure 16). Thus far, we have not managed to solve its crystal structure by using the direct space methodology, being an additional difficulty the presence of two different BL molecules in the asymmetric unit. Further work is still in progress.

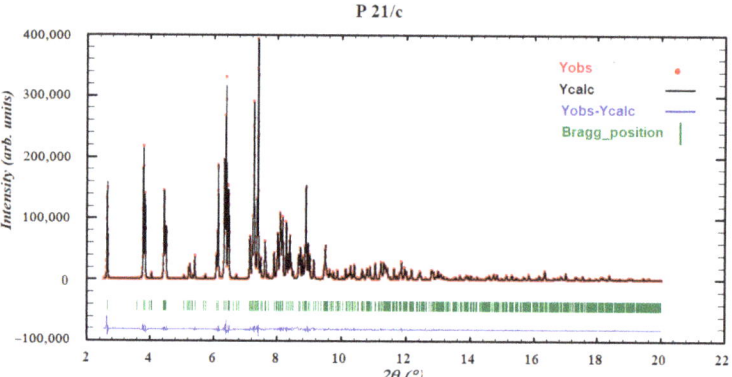

Figure 16. Pawley pattern matching plot of form II. Agreement factor: Rwp = 1.3%. The plot shows the experimental powder XRD profile (red+marks), the calculated powder XRD profile (black solid line) and the difference profile (blue, lower line). Tick marks indicate peak positions.

3.5.2. Bilastine Chloroform Solvates

As it has been mentioned, the solvated phase S$_{3CHCl3\cdot H2O}$ has resulted to be unstable, the crystals losing solvent rapidly after removal from the mother liquor and generally rendering manipulation of the material and accurate analysis of its composition impossible. The first visible signs of desolvation, such as opacification of the crystals, were observed only after seconds at r.t. However, its crystal structure has been successfully determined at 100K resulting in an orthorhombic $Pna2_1$ cell with Z = 4. In S$_{3CHCl3\cdot H2O}$ BL molecules are linked by hydrogen bonds involving the carboxylate oxygen of one BL molecule and the piperidinium nitrogen of another BL in an alternated mode similar to form I, forming corrugated layer structures (Figure 17).

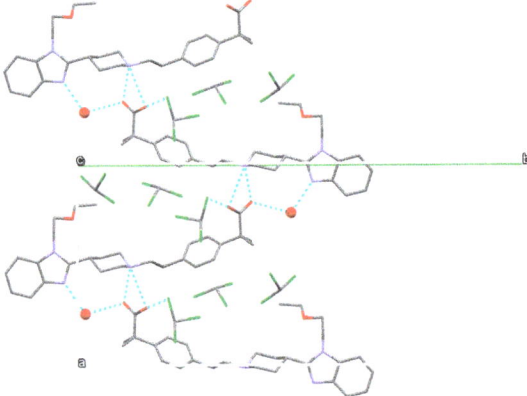

Figure 17. Carboxylate/piperidinium H-bond interactions between BL molecules observed in the crystal structure of S$_{3CHCl3\cdot H2O}$. Water molecules represented by red balls. Hydrogens have been omitted for clarity.

Water and chloroform molecules occupy voids inside the crystal structure as it is shown in Figure 18.

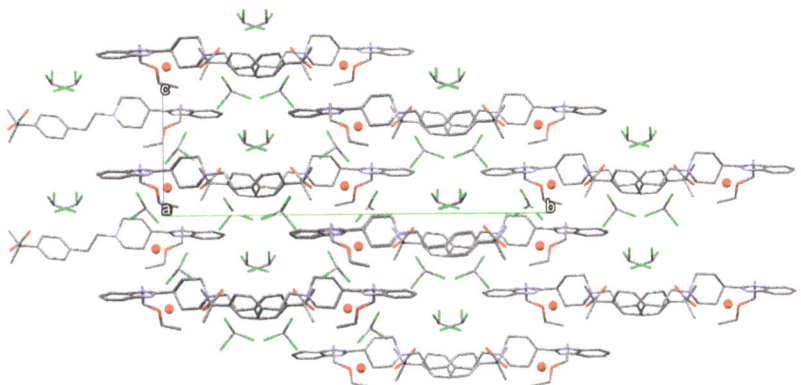

Figure 18. Packing diagram of form S$_{3CHCl3-H2O}$ viewed parallel to [100]. Water molecules represented by red balls. Hydrogens have been omitted for clarity.

The X-ray structural analysis reveals the presence of one hydrogen-bonded water molecule as well as three hydrogen- and halogen-bonded chloroform molecules and one BL in its asymmetric unit. As shown in Figure 19, the water molecule acts as hydrogen bond donor in a DD environment, its hydrogens showing H-bonding with the nitrogen of the benzimidazol of one BL, on one hand (distance: 2.077 Å), and with an oxygen of the carboxylate of another BL molecule, on the other hand (distance: 2.133 Å). This carboxylate oxygen is also involved in an asymmetric three-center (or bifurcated) hydrogen bond system (distances: 1.887 Å and 2.452 Å) with the piperidinium of the first BL (being the structure zwitterionic), forming altogether R^2_3(11) and R^2_1(4) supramolecular heterosynthons. Moreover, one chloroform molecule interacts via halogen bond with the other oxygen of the carboxylate, being 3.139 Å its Cl$_2$HC-Cl···O=C distance, shorter than the sum of the van der Waals radii of the involved atoms. Moreover, this chloroform molecule is involved in a Cl$_3$C-H···O=C (distance: 2.142 Å) hydrogen bond with the carboxylate and in a Cl$_2$HC-Cl···Cl-CHCl$_2$ (distance: 3.366 Å) halogen bond with a second chloroform molecule which in turn interacts with the third chloroform forming also a Cl$_2$HC-Cl···Cl-CHCl$_2$ (distance: 3.322 Å) halogen bond. The second chloroform is involved also in a Cl$_3$C-H···π (distance Ph-center: 2.354 Å) interaction with the phenyl ring of another BL and the third chloroform in turn interacts also with the water molecule via Cl$_3$C-H···O-H$_2$ (distance: 2.011 Å) hydrogen bond.

As stated, it was not possible to grow single crystals of S$_{CHCl3-H2O}$ as once its precursor solvate was obtained, the crystals were not stable and two chloroform molecules were rapidly lost after removal from the mother liquor, leaving a darkened crystal. Therefore, its powder diffraction pattern was indexed to an orthorhombic cell and the space group perfectly determined to be *Pna*2$_1$ from the systematic absences. The crystal structure was solved with the direct-space strategy TALP, and its refinement was performed by the Rietveld method. Figure 20 depicts the final Rietveld plot.

The X-ray structural analysis of the structure of S$_{CHCl3-H2O}$ solved from PXRD reveals the presence of one hydrogen-bonded water molecule, one hydrogen-bonded chloroform molecule and one BL in its asymmetric unit. BL molecules are also linked by hydrogen bonds involving the carboxylate oxygen of one BL molecule and the piperidinium nitrogen of another BL in an asymmetric three-center hydrogen bond system (distances: 1.731 Å and 2.253 Å), with the same R^2_1(4) heterosynthon, forming corrugated layer structures as in S$_{3CHCl3-H2O}$. In this case, an interlayer interaction is accomplished by π···π parallel displaced phenyl stacking of the alternated BL (observed lateral offset stack: 4.66 Å, centroid to centroid distance d($\pi \cdots \pi$): 5.19 Å, interplanar distance: 2.29 Å). Moreover, the water molecule acts as hydrogen bond donor in a D environment as only one hydrogen is involved in hydrogen-bonding (distance: 1.942 Å) with the nitrogen of the benzimidazol of one BL. Finally, the chloroform molecule is involved in a $Cl_3C-H \cdots O=C$ (distance: 2.337 Å) hydrogen bond with the carboxylate (Figure 21).

Unlike solvate $S_{3CHCl3-H2O}$, the water molecule cannot form another H-bond with an oxygen of the carboxylate of another BL molecule as the carboxylate is not pointing towards it as it has rotated some degrees being in this case too far. The conformations of the BL molecules in both heterosolvates are compared in Figure 22, differing particularly with respect to the orientation of the carboxylate fragment.

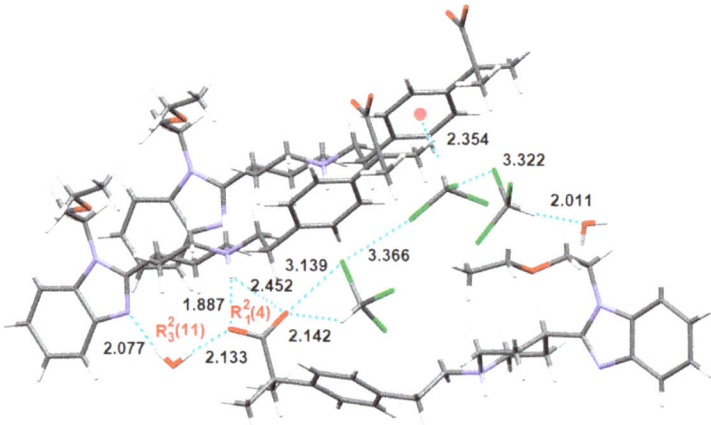

Figure 19. Intermolecular interactions (distances in Å) observed in the crystal structure of $S_{3CHCl3-H2O}$.

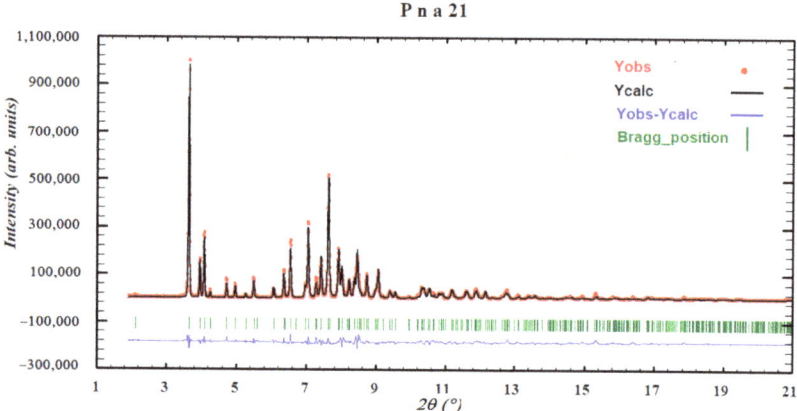

Figure 20. Final Rietveld plot for the crystal structure refinement of form $S_{CHCl3-H2O}$. Agreement factor: Rwp = 2.3%. The plot shows the experimental powder XRD profile (red+marks), the calculated powder XRD profile (black solid line) and the difference profile (blue, lower line). Tick marks indicate peak positions.

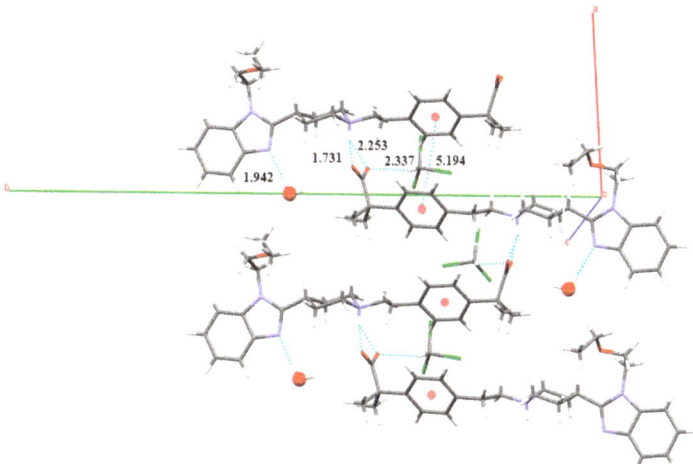

Figure 21. Intermolecular interactions (distances in Å) observed in the crystal structure of $S_{CHCl3-H2O}$.

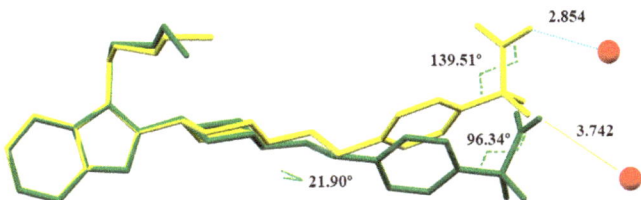

Figure 22. Overlay of BL molecules of form $S_{3CHCl3-H2O}$ (yellow) and $S_{CHCl3-H2O}$ (green) showing that the water molecule (represented by red balls) is too far (distances in Å) from the carboxylate of $S_{CHCl3-H2O}$. Hydrogen atoms are omitted for clarity.

The channeled crystal structure of solvate $S_{CHCl3-H2O}$ is shown in Figure 23 where chloroform and water molecules occupy different channels.

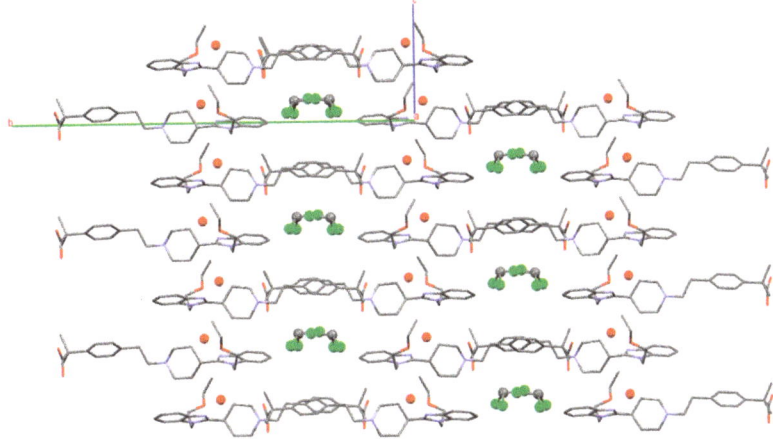

Figure 23. Crystal packing of form $S_{CHCl3-H2O}$ viewed parallel to [100]. Water molecules represented by red balls. Hydrogens have been omitted for clarity.

Finally, single crystals of S$_{CHCl3}$ were grown in anhydrous chloroform. This solvate crystallizes in the monoclinic space group $P2_1$ with two molecules of BL and two molecules of chloroform in the asymmetric unit. Again, BL molecules are linked by hydrogen bonds (distance: 1.583 Å) involving one oxygen of the carboxylate and the piperidinium NH, forming cascade layers as in Form III (Figure 24).

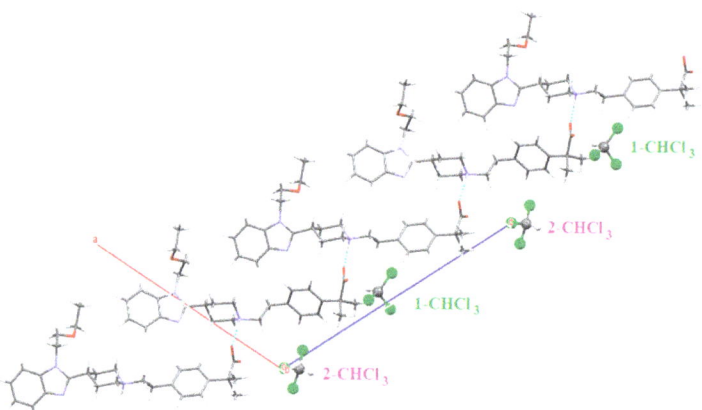

Figure 24. Hydrogen bonds between BL molecules observed in form S$_{CHCl3}$.

There are two different chloroform molecules in this solvate: 1-CHCl$_3$ which is involved in a Cl$_3$C-H···O=C (distance: 2.008 Å) hydrogen bond with the carboxylate of one BL and 2-CHCl$_3$ which forms the same H-bond (distance: 2.013 Å) with the carboxylate of the other BL (Figure 25). This chloroform participates also in a Cl···π interaction with the benzene ring of another BL (Cl-centroid distance: 3.683 Å, C-Cl-centroid angle: 159.51°). Figure 26 shows the crystal packing of S$_{CHCl3}$.

The crystal data of all the crystal structures solved for BL are summarized in Table S15 (from SCXRD) and Table S16 (from PXRD) of the Supplementary Material.

The three anhydrous polymorphs of BL crystallize in the same space group $P2_1/c$ and Form II shows a unit cell similar to Form III except for the parameter "a" which is approximately twice as much as in Form III. This results in a cell for Form II with double volume than forms I and III and with Z = 8.

On the other hand, both heterosolvates S$_{3CHCl3-H2O}$ and S$_{CHCl3-H2O}$ crystallize in the same orthorhombic space group $Pna2_1$ showing similar unit cell parameters, while the monosolvate S$_{CHCl3}$ crystallizes in $P2_1$, the three solvates with Z = 4.

BL is a molecule which shows an imbalance between hydrogen bond donors and acceptor groups, having only one N-H donor group. All the structures solved are zwitterionic and BL molecules are linked by hydrogen bonds involving the carboxylate oxygen and the piperidinium nitrogen. The incorporation of water and chloroform molecules in the solvates of BL can be attributed to the imbalance in the ratio of donors and acceptors groups. Water molecules act as hydrogen bond donors in both heterosolvates, interacting with the nitrogen of the benzimidazol of BL. Instead, chloroform molecules form hydrogen bonds with the carboxylate in all solvates. Some halogen bonds are also observed in these structures, confirming that halogen bonding is one of the stabilizing interactions in chloroform solvates. However, as it has been stated [30], halogen bonding does not seem to be sufficiently strong to retain the solvent in the crystals on its own as other short contacts are also present. In fact, the two chloroform molecules in S$_{3CHCl3-H2O}$ which are not linked by hydrogen bonds to the carboxylate, are lost when this solvate is removed from the solution. On the other hand, a water molecule captured from the air is incorporated in the crystal structure of S$_{CHCl3}$ when taking it out from the solution, probably to further stabilize its structure. In fact, S$_{CHCl3}$ can be considered a hygroscopic solid form as it has the ability to take up and retain water vapor. This is also true for many other compounds such as 1,10-phenanthroline [27].

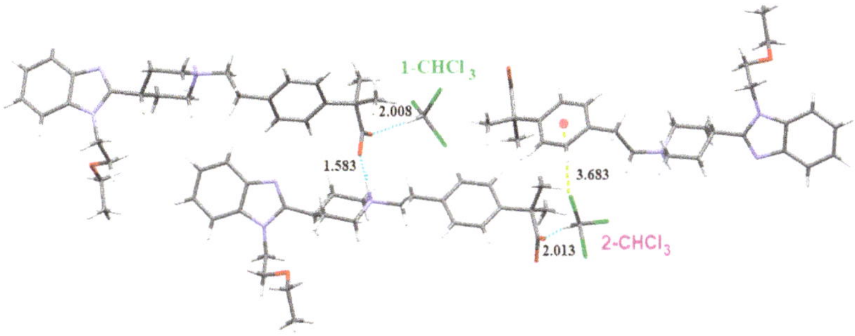

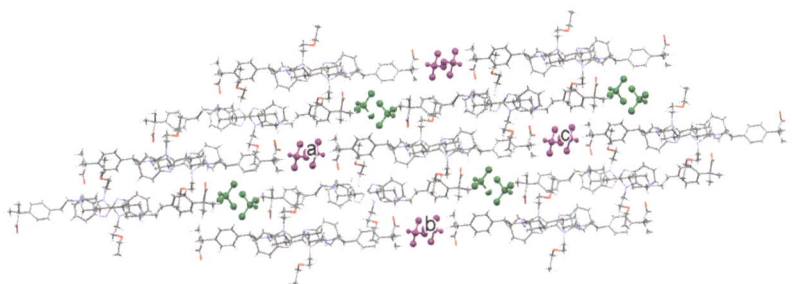

Figure 25. Intermolecular interactions (distances in Å) observed in solvate S_{CHCl3}.

Figure 26. Crystal packing of S_{CHCl3} along the b axis. Green color for 1-CHCl$_3$ and purple color for 2-CHCl$_3$.

4. Conclusions

Knowledge of even highly unstable and transient solvates is crucial for understanding the formation of a specific solid form and for developing pharmaceuticals and needs to be considered. Solvates that desolvate readily as soon as they are removed from the mother liquor can easily be overlooked in polymorph screening programs, and wrong conclusions concerning the solvent effects on the nucleation of individual solid forms may result. The current study clearly indicates the existence of two elusive chloroform solvates of BL, the heterosolvate $S_{3CHCl3\cdot H2O}$ and the monosolvate S_{CHCl3}, both transforming into solvate $S_{CHCl3\cdot H2O}$ when being exposed to ambient conditions, outside the solution of crystallization. Thus, three chloroform solvates of BL with diverse stoichiometry have been discovered. Solvate formation may be easily rationalized based on a mismatch of H-bonding donor or acceptor groups. The reason for solvate formation in BL is the lack of H-bonding donor groups. The analyses of their crystal structures when comparing with the anhydrous forms shows that $S_{CHCl3\cdot H2O}$, the more stable solvate, incorporates one water molecule and one chloroform molecule to satisfy the acceptor capacities of the benzimidazole and the carboxylate, respectively. Thus, this study adds more data about the structural reasons for hydrate formation. Interesting and novel, solvate $S_{CHCl3\cdot H2O}$ can serve as a useful intermediate towards any of the three anhydrous forms of BL, provided the right conditions are chosen.

Supplementary Materials: The following are available online at https://www.mdpi.com/article/10.3390/cryst11040342/s1. Details of the solid form screening are shown in Tables S1–S14. Tables S15 and S16 summarize the crystal data of all the crystal structures solved in the present study for BL from SCXRD and from PXRD, respectively.

Author Contributions: Conceptualization, C.P.; methodology, C.P., A.P. and A.B.; software, C.P., M.F.-B. and O.V.; validation, C.P., A.P. and A.B.; formal analysis, C.P., A.P., A.B., M.F.-B. and O.V.; investigation, C.P., A.P. and A.B.; data curation, C.P., A.P. and A.B.; writing—original draft preparation, C.P.; writing—review and editing, C.P.; supervision, C.P. All authors have read and agreed to the published version of the manuscript.

Funding: This research received no external funding.

Acknowledgments: ALBA synchrotron is acknowledged for the provision of beamtime.

Conflicts of Interest: The authors declare no conflicts of interest.

References

1. Singhal, D.; Curatolo, W. Drug polymorphism and dosage form design: A practical perspective. *Adv. Drug Deliv. Rev.* **2004**, *56*, 335–347. [CrossRef]
2. Bladgen, N.; Davey, R.J. Polymorph Selection: Challenges for the Future? *Cryst. Growth Des.* **2003**, *3*, 873–885.
3. Zalte, A.G.; Darekar, A.B.; Gondkar, S.B.; Saudagar, R.B. Cocrystals: An Emerging Approach to Modify Physicochemical Properties of Drugs. *Am. J. PharmTech Res.* **2014**, *4*, 2056–2072.
4. Rodriguez-Spong, B.; Price, C.P.; Jayasankar, A.; Matzger, A.J.; Rodriguez-Hornedo, N. General principles of pharmaceutical solid polymorphism: A supramolecular perspective. *Adv. Drug Delivery Rev.* **2004**, *56*, 241–274. [CrossRef]
5. Grodowska, K.; Parezewski, A. Organic solvents in the pharmaceutical industry. *Acta Pol. Pharm. Drug Res.* **2010**, *67*, 3–12.
6. Zvoníček, V.; Skořepová, E.; Dušek, M.; Babor, M.; Žvátora, P.; Šoóš, M. First crystal structures of pharmaceutical ibrutinib: Systematic solvate screening and characterization. *Cryst. Growth Des.* **2017**, *17*, 3116–3127. [CrossRef]
7. Wang, Y.; Chi, Y.; Zhang, W.; Yang, Q.; Yang, S.; Su, C.; Lin, Z.; Gu, J.; Hu, C. Structural diversity of diosgenin hydrates: Effect of initial concentration, water volume fraction, and solvent on crystallization. *Cryst. Growth Des.* **2016**, *16*, 1492–1501. [CrossRef]
8. Berzins, A.; Trimdale, A.; Kons, A.; Zvaniņa, D. On the formation and desolvation mechanism of organic molecule solvates: A structural study of methyl cholate solvates. *Cryst. Growth Des.* **2017**, *17*, 5712–5724. [CrossRef]
9. Tieger, E.; Kiss, V.; Pokol, G.; Finta, Z.; Rohlíček, J.; Skořepová, E.; Dušek, M. Rationalization of the formation and stability of bosutinib solvated forms. *CrystEngComm* **2016**, *18*, 9260–9274. [CrossRef]
10. Council of Europe. Residual Solvents. In *European Pharmacopoeia*; Council of Europe: Strasbourg, France, 2013; p. 5967.
11. Morissette, S.L.; Almarsson, O.; Peterson, M.L.; Remenar, J.F.; Read, M.J.; Lemmo, A.V.; Ellis, S.; Cima, M.J.; Gardner, C.R. High-throughput crystallization: Polymorphs, salts, co-crystals and solvates of pharmaceutical solids. *Adv. Drug Deliv. Rev.* **2004**, *56*, 275–300. [CrossRef] [PubMed]
12. Wolthers, O.D. Bilastine: A new nonsedating oral H1 antihistamine for treatment of allergic rhinoconjunctivitis and urticaria. *BioMed Res. Int.* **2013**, *2013*, 1–6. [CrossRef]
13. Corcóstegui, R.; Labeaga, L.; Innerárity, A.; Berisa, A.; Orjales, A. Preclinical pharmacology of bilastine, a new selective histamine H 1 receptor antagonist. *Drugs R D* **2005**, *6*, 371–384. [CrossRef] [PubMed]
14. Orjales, A.; Rubio, V.; Bordell, M. Benzimidazole Derivatives with Antihistaminic Activity. EP 0818454B1, 3 June 1997.
15. Orjales, A.; Bordell, M.; Canal, G.; Blanco, H. Polymorph of Acid 4-[2-[4-[1-(2-Ethoxyethyl)-1h-Benzimidazole-2-Il]-1-Piperidinyl]Ethyl]-$G(A), $G(A)-Dimethyl-Benzeneacetic. WO 2003/089425, 30 October 2003.
16. Fauth, F.; Peral, I.; Popescu, C.; Knapp, M. The new Material Science Powder Diffraction beamline at ALBA Synchrotron. *Powder Diffr.* **2013**, *28*, S360–S370. [CrossRef]
17. Boultif, A.; Louër, D.J. Powder pattern indexing with the dichotomy method. *J. Appl. Crystallogr.* **2004**, *37*, 724–731. [CrossRef]
18. Vallcorba, O.; Rius, J.; Frontera, C.; Peral, I.; Miravitlles, C. DAJUST: A suite of computer programs for pattern matching, space-group determination and intensity extraction from powder diffraction data. *J. Appl. Crystallogr.* **2012**, *45*, 844–848. [CrossRef]
19. Vallcorba, O.; Rius, J.; Frontera, C.; Miravitlles, C. TALP: A multisolution direct-space strategy for solving molecular crystals from powder diffraction data based on restrained least squares. *J. Appl. Crystallogr.* **2012**, *45*, 1270–1277. [CrossRef]
20. Sheldrick, G.M. A Short History of SHELX. *Acta Crystallogr.* **2008**, *A64*, 112–122. [CrossRef] [PubMed]
21. Hübschle, C.B.; Sheldrick, G.M.; Dittrich, B.J. ShelXle: A Qt graphical user interface for SHELXL. *J. Appl. Crystallogr.* **2011**, *44*, 1281–1284. [CrossRef]
22. Burger, A.; Ramberger, R. On the polymorphism of pharmaceuticals and other molecular crystals. I. *Mikrochim. Acta* **1979**, 259–271. [CrossRef]
23. Yu, L.J. Inferring thermodynamic stability relationship of polymorphs from melting data. *J. Pharm. Sci.* **1995**, *84*, 966–974. [CrossRef]
24. Haleblian, J.; Crone, W.M. Pharmaceutical applications of polymorphism. *J. Pharm. Sci.* **1969**, *58*, 911–929. [CrossRef]
25. Zencirci, N.; Griesser, U.J.; Gelbrich, T.; Kahlenberg, V.; Jetti, R.K.R.; Apperley, D.C.; Harris, R.K. New solvates of an old drug compound (phenobarbital): Structure and stability. *J. Phys. Chem. B* **2014**, *118*, 3267–3280. [CrossRef] [PubMed]
26. Ward, M.R.; Oswald, D.H. Hidden solvates and transient forms of trimesic acid. *Crystals* **2020**, *10*, 1098. [CrossRef]

27. Braun, D.E.; Schneeberger, A.; Griesser, U.J. Understanding the role of water in 1, 10-phenanthroline monohydrate. *CrystEngComm* **2017**, *19*, 6133–6145. [CrossRef] [PubMed]
28. Petkune, S.; Bobrovs, R.; Actins, A. Organic solvents vapor pressure and relative humidity effects on the phase transition rate of α and β forms of tegafur. *Pharm. Dev. Technol.* **2012**, *17*, 625–631. [CrossRef] [PubMed]
29. Hunter, C.A.; Sanders, J.K.M. The nature of. pi.-. pi. interactions. *J. Am. Chem. Soc.* **1990**, *112*, 5525–5534. [CrossRef]
30. Takieddin, K.; Zhimyak, Y.Z.; Fábián, L. Prediction of hydrate and solvate formation using statistical models. *Cryst. Growth Des.* **2016**, *16*, 70–81. [CrossRef]

Article

Improving the Solubility of Aripiprazole by Multicomponent Crystallization

Qi Zhou [1,2], Zhongchuan Tan [1,2], Desen Yang [1,2], Jiyuan Tu [1,2], Yezi Wang [1,2], Ying Zhang [1,2], Yanju Liu [1,2,*] and Guoping Gan [1,2,*]

[1] Pharmacy Faculty, Hubei University of Chinese Medicine, Wuhan 430070, China; zhouqi0908@126.com (Q.Z.); tanzhongchuan2021@163.com (Z.T.); yds22@126.com (D.Y.); 3070@hbtcm.edu.cn (J.T.); WYZworkmail@163.com (Y.W.); zy09242021@163.com (Y.Z.)

[2] Technical Engineering Research Center of Traditional Chinese Medicine Processing in Hubei Province, Wuhan 430070, China

* Correspondence: lyj1965954@hbtcm.edu.cn (Y.L.); ganguop@sina.com (G.G.)

Abstract: Aripiprazole (ARI) is a third-generation antipsychotic with few side effects but a poor solubility. Salt formation, as one common form of multicomponent crystals, is an effective strategy to improve pharmacokinetic profiles. In this work, a new ARI salt with adipic acid (ADI) and its acetone hemisolvate were obtained successfully, along with a known ARI salt with salicylic acid (SAL). Their comprehensive characterizations were conducted using X-ray diffraction and differential scanning calorimetry. The crystal structures of the ARI-ADI salt acetone hemisolvate and ARI-SAL salt were elucidated by single-crystal X-ray diffraction for the first time, demonstrating the proton transfer from a carboxyl group of acid to ARI piperazine. Theoretical calculations were also performed on weak interactions. Moreover, comparative studies on pharmaceutical properties, including powder hygroscopicity, stability, solubility, and the intrinsic dissolution rate, were carried out. The results indicated that the solubility and intrinsic dissolution rate of the ARI-ADI salt and its acetone hemisolvate significantly improved, clearly outperforming that of the ARI-SAL salt and the untreated ARI. The study presented one potential alternative salt of aripiprazole and provided a potential strategy to increase the solubility of poorly water-soluble drugs.

Keywords: aripiprazole; multicomponent crystal; crystal structure; solubility

Citation: Zhou, Q.; Tan, Z.; Yang, D.; Tu, J.; Wang, Y.; Zhang, Y.; Liu, Y.; Gan, G. Improving the Solubility of Aripiprazole by Multicomponent Crystallization. *Crystals* **2021**, *11*, 343. https://doi.org/10.3390/cryst11040343

Academic Editor: Emilio Parisini

Received: 27 February 2021
Accepted: 25 March 2021
Published: 28 March 2021

Publisher's Note: MDPI stays neutral with regard to jurisdictional claims in published maps and institutional affiliations.

Copyright: © 2021 by the authors. Licensee MDPI, Basel, Switzerland. This article is an open access article distributed under the terms and conditions of the Creative Commons Attribution (CC BY) license (https://creativecommons.org/licenses/by/4.0/).

1. Introduction

In the past few decades, a number of drugs have shown poor physicochemical properties, especially aqueous solubility and stability, often affecting their absorption in the gastrointestinal tract [1]. Multicomponent crystal formation is an effective strategy to improve pharmacokinetic profiles without altering the main chemical structures and inherent biological activity [2,3]. Co-crystals and salts are two common forms of multicomponent crystals which might have higher solubility and faster dissolution behavior compared to untreated drugs [4,5]. Compared to cocrystals, salt formation is the simplest and most cost-effective strategy and has significant advantages in addressing poor aqueous solubility because of ionizable drugs [6].

Aripiprazole (ARI), a third-generation antipsychotic, is a dopamine D2 receptor partial agonist and D1 receptor agonist which can ameliorate hyperprolactinemia induced by other antipsychotic drugs and cause fewer side effects, such as weight gain, diabetes, and dyslipidemia [7,8]. However, ARI belongs to the Biopharmaceutics Classification System (BCS) class II and its clinical use is limited by its poor aqueous solubility [9,10]. As a weakly basic drug, salt formation is one of the most popular and effective approaches to improve physicochemical properties, especially solubility [6,11,12]. Many salts of ARI have been reported. Freire et al. reported successively eight crystal structures of ARI salts with nitrate, perchlorate, oxalate, phthalate, homophthalate, thiosalicilate, and two different dihidrogenphosphates [13–16]. Nanubolu et al. reported five ARI salts with benzoic acid, 2,4-dihydroxy benzoic acid, 2,5-dihydroxy benzoic acid, salicylic acid, and hydrochloric acid [17]. However, these studies mainly focus on the structure illustration, and key pharmaceutical properties were not given. While Zhao et al. synthesized six ARI salts with gallic acid, 4-aminosalicylic acid, acetylsalicylic acid, maleic acid, fumaric acid, and malic acid and evaluated their solubility and dissolution profile [18], new ARI salts with ideal pharmaceutical properties are still in demand.

Continuing to explore the excellent salts of ARI, we successfully obtained a new ARI salt with adipic acid (ADI) and its acetone hemisolvate, along with a known ARI salt with salicylic acid (SAL). The molecular structures of the ARI and cocrystal formers (CCF) are displayed in Scheme 1. Their comprehensive characterizations were conducted using powder X-ray diffraction (PXRD) and differential scanning calorimetry (DSC). The crystal structures of the ARI-ADI salt acetone hemisolvate and ARI-SAL salt were elucidated by single-crystal X-ray diffraction (SXRD) for the first time. Furthermore, computational studies, including molecular electrostatic potential surface (MEPS) and Hirshfeld surface analysis (HSA), were applied to explore molecular interactions between the active pharmaceutical ingredient (API) and CCF. Above all, their pharmaceutical properties, such as powder hygroscopicity, stability, solubility, and the intrinsic dissolution rate (IDR), were evaluated. The results showed that the solubility and IDR of the ARI-ADI salt and its acetone hemisolvate significantly improved. This study provides a potential strategy to increase the solubility of poorly water-soluble drugs and gain a comprehensive understanding of the structure–property relationship.

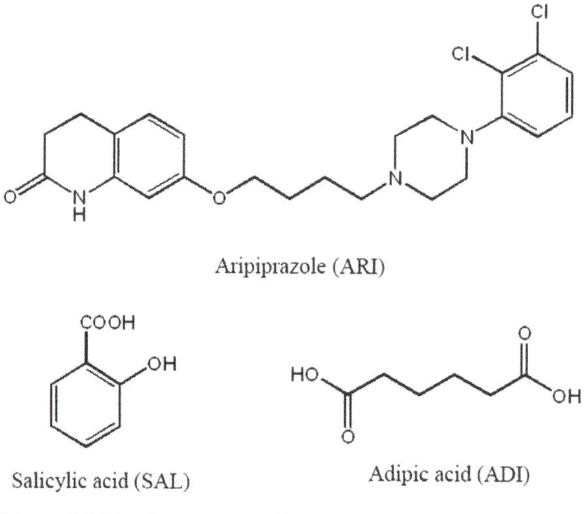

Scheme 1. Molecular structures of aripiprazole and cocrystal formers.

2. Materials and Methods

2.1. Materials

The aripiprazole was purchased from Sun Chemical & Technology (Shanghai, China) Co., Ltd., and the CCF were purchased from J&K Scientific Ltd. (Beijing, China). All chemicals were used without further purification. All reagents used for this study were of analytical grade and purchased from Sinopharm Chemical Reagent Co., Ltd. (Shanghai, China). Purified water was prepared by Millipore UP Water Purification System (Merck, Kenilworth, NJ, USA).

2.2. Preparation of Multicomponent Crystals

ARI-SAL salt. The equimolar amounts of ARI (224.2 mg, 0.5 mmol) and SAL (69.1 mg, 0.5 mmol) were dissolved in 30 mL of acetonitrile-water mixed solvent (v/v, 1:1) at 60 °C. The resulting solution was filtered and then kept for eight days at 35 °C. White block crystals were obtained. The bulk samples were prepared by slurry method. A mixture of ARI (448 mg, 1 mmol) and SAL (138 mg, 1 mmol) was added to 6 mL of acetonitrile-water mixed solvent (v/v, 1:1) and stirred for 12 h at a speed of 350 rpm. The solid obtained by filtering the solution was dried to a constant weight in a vacuum oven at 30 °C.

ARI-ADI salt acetone hemisolvate. The mixture of ARI (224.2 mg, 0.5 mmol) and ADI (36.5 mg, 0.25 mmol) was dissolved in 20 mL of acetone at 60 °C. The resulting solution was filtered and then kept for four days at 35 °C. White block crystals were obtained. The bulk samples were prepared by slurry method. A mixture of ARI (448 mg, 1 mmol) and ADI (73 mg, 0.5 mmol) was added into 6 mL of acetone solution and stirred for 12 h at a speed of 350 rpm. The solid obtained by filtering the solution was dried to a constant weight in a vacuum oven at 30 °C.

Additionally, ARI-ADI salt was obtained by drying ARI-ADI salt acetone hemisolvate in a vacuum oven at 70 °C for 2 h.

2.3. Characterization

2.3.1. Single Crystal X-ray Diffraction (SXRD)

The crystal structures were determined with a Bruker APEX-II CCD diffractometer using graphite monochromatic Mo-Kα radiation (λ = 0.71073 Å) at 296K. Frame integration was performed using SAINT (version 7.68A) [19]. The resulting raw data were scaled and the absorption was corrected using a multi-scan averaging of symmetry-equivalent data by SADABS [20]. The structure was solved by the direct method with the olex2 software and then refined via full-matrix least-squares procedures using SHELXL-2014 on F2 with anisotropic displacement parameters (ADPs) for non-hydrogen atoms [21,22]. The H atoms were located in idealized difference Fourier maps and refined as riding models with isotropic thermal parameters (Uiso(H) = 1.2 Ueq(C), Uiso(H) = 1.2 Ueq(N), and Uiso(H) = 1.5 Ueq(O)).

Diamond [23] (version 4.6.3, Crystal Impact GbR, Bonn, Germany) was used for preparing crystal packing diagrams.

2.3.2. Powder X-ray Diffraction (PXRD)

Powder X-ray diffraction (PXRD) patterns were recorded on a Bruker D8 ADVANCE X-ray powder diffractometer (Bruker, Germany) with Cu-Kα radiation at 40 KV and 40 mA. After sieving through 100 mesh, about 50 mg samples were measured in the 2-theta range of 5–50° at a scan rate of 8°/min.

2.3.3. Differential Scanning Calorimetry (DSC)

DSC analyses were taken on a Mettler Toledo DSC1 instrument (Mettler, Zurich, Switzerland). An amount of 3~5 mg samples were put into aluminum pans with pinhole lids and heated in the 30~300 °C temperature range at a constant heating rate of 10 °C/min under a nitrogen flux of 50 cm^3/min.

2.4. Computational Studies

2.4.1. Acid Dissociation Constant (pKa)

The formation of the resultant supramolecules can be predicted according to the pKa difference (ΔpKa = pKa[base] − pKa[acid]) of corresponding acid/base pairs [24,25]. It is generally accepted that a salt will be formed if the ΔpKa value is greater than 3, and a ΔpKa value less than 0 will lead to the formation of cocrystals. However, if the ΔpKa value is in the range of 0~3, it will be not accurate to predict the resulting formation [25]. The pKa values of ARI and CCF were calculated by MarvinSketch 15.6.29 (ChemAxon, Budapest, Hungary) [26].

2.4.2. Molecular Electrostatic Potential Surface (MEPS)

The molecular structures of ARI and CCF were extracted from their crystal structures. Full geometry optimization and wave functions were performed by density function theory (DFT) using the B3LYP hybrid functional with 6-311G(d) basis set in the Gaussian 09 software. MEPS was mapped on the 0.001 a.u. electron density isosurface and analyzed by the Multiwfn program [27] and VWD program [28].

2.4.3. Hirshfeld Surface Analysis (HSA)

HSA and fingerprint plots were performed by CrystalExplorer 17.5 software [29], providing information about the nature of intermolecular interactions and their quantitative contribution to the Hirshfeld surface [30].

2.5. Powder Hygroscopicity

Dry glass weighing bottles (outside diameter 50 mm, height 25 mm) were placed in a dryer with ammonium chloride saturated solution at 25 °C ± 1 °C and weighed (m1) after 24 h of storage. Then, about 100 mg samples were put into the bottles and weighed accurately (m2), respectively. After storage at the aforementioned conditions with the caps opened for 24 h, each sample was weighed again (m3). The hygroscopicity could be calculated by the following equation.

$$\text{Mass Change (\%)} = (m3 - m1) / (m2 - m1) \times 100\%.$$

2.6. Stability Test

The stabilities of ARI-SAL salt and ARI-ADI salt acetone hemisolvate were studied under high-temperature (60 ± 1 °C) and high-humidity (95 ± 5%) tests. Materials were randomly selected and exposed to high temperature and high humidity conditions. The storage times were 10 days and PXRD was applied to measure the final samples.

2.7. Solubility Experiment

2.7.1. Solubility Test

The solubility measurements were performed under water (pH 7.0), hydrochloric acid buffer (pH 1.2), acetate acid-sodium acetate buffered solution (pH 4.5), and phosphate buffer solution (pH 6.8), respectively. An excess number of samples were added to a test tube containing 10 mL of solvent and the tube was shaken in an orbital shaker (37 ± 0.5 °C) until reaching the equilibrium condition (48 h). The solution was filtered through a 0.45 μm mixed cellulose ester membrane and then analyzed by HPLC (LC-20, Shimadzu, Japan) equipped with a SP-20 ultraviolet detector at 254 nm wavelength. Quantitative tests were performed on a YMC C8 column (250 × 4.6 mm, 5 μm) with an external standard method. The mobile phase consisted of acetonitrile and water (containing 0.1% trimethylamine) (65:35, v/v), with a flow rate of 1.0 mL/min. Each test was performed in triplicate.

2.7.2. Intrinsic Dissolution Rate (IDR) Test

IDR is a key physicochemical parameter commonly used to assess in vivo dissolution and reflect bioavailability of drugs [31]. In this work, IDR was measured by the rotary basket method, which was applied to distinguish their dissolution properties. Prior to the IDR test, round discs of the samples should be compressed with a hydraulic press (Jintan Ruiding Machinery Co., Ltd. Changzhou, China). Specifically, 300 mg samples were compressed at a pressure of 115.2 MPa for 10 s to form smooth discs 8 mm in diameter. The acquired discs were coated with beeswax on three sides. The intrinsic dissolution study was performed at 100 rpm in 500 mL of hydrochloric acid buffer (pH 1.2) as a dissolution medium at 37 °C. The ARI concentration in solution was measured at the predetermined time interval by the same analysis method as the solubility test. The sink conditions were maintained during the entire dissolution experiment, and each test was performed in triplicate.

3. Results and Discussion

3.1. Characterization

3.1.1. SXRD Analysis

The crystal structures revealed that both ARI-SAL salt and ARI-ADI salt acetone hemisolvate were salts and the protonation occurred at the N_2 atom, forming a strong charge assisted hydrogen bond formed by the strongest acceptor from the carboxylate anion interacting with strongest N^+-H donor of piperazinium (see Appendix A). The crystallographic data are listed in Table 1 and the hydrogen bonding parameters are given in Table 2. The C_{16}-$H_{16A}\cdots Cl_1$ hydrogen bonds in Table 2 correspond to the typical intramolecular interaction characteristic of the dichlorophenyl-1-piperazinyl group in all reported ARI variants [16–18].

Table 1. Crystallographic data of two multicomponent crystals of aripiprazole.

Compound Reference	ARI-SAL Salt	ARI-ADI Salts Acetone Hemisolvate
Chemical formula	$C_{23}H_{28}Cl_2N_3O_2 \cdot C_7H_5O_3$	$C_{23}H_{28}Cl_2N_3O_2 \cdot 0.5(C_6H_8O_4) \cdot 0.5(C_3H_6O)$
Formula mass	586.49	550.48
Crystal system	monoclinic	triclinic
Space group	$P\,2_1/c$	P-1
a/Å	15.082 (2)	7.6388(12)
b/Å	9.6912 (13)	10.7268(17)
c/Å	21.220 (3)	18.390(3)
α/°	90	97.229(3)
β/°	106.743 (2)	93.641(3)
γ/°	90	105.529(3)
Unit cell volume/Å^3	2970.1 (7)	1432.9(4)
Temperature/k	296(2)	296(2)
No. of formula units per unit cell, Z	4	2
Crystal density (g/cm^3)	1.312	1.276
No. of reflections measured	9612	7725
No. of independent reflections	6483	4047
R_{int}	0.0289	0.0321
Final R_1 values (I > 2σ(I))	0.0806	0.0689
Final w_R (F2) values (I > 2σ(I))	0.2526	0.2135
Final R_1 values (all data)	0.1127	0.1308
Final w_R (F2) values (all data)	0.2848	0.2725
F(000)	1232	582
Goodness of fit on F2	1.084	0.963
CCDC Number	1991810	2023732

ARI-SAL salt. The ARI-SAL salt crystallized in the monoclinic $P2_1/c$ space group with an aripiprazole cation, balanced by a salicylic counter-ion, in the asymmetric unit (Figure 1a). The O_5 atom of salicylic anion presented positional disorder over two sites with a 0.6: 0.4 site-occupancy. ARI-SAL salt, like other lactam compounds, formed a centrosymmetric N_1-$H_1\cdots O_1$ dimer $R_2^2(8)$ motif with its inversion-related molecule [32]. Interestingly, the dimer formed a 1D ribbon through the short $Cl_1\cdots O_1$ interaction, which formed another new dimeric substructure (Figure 2a). Furthermore, the dimer units were arranged helically into the three-dimensional network (Figure 2b) by the bifurcated N-H$\cdots$O hydrogen bonds (N_2^+-$H_2\cdots O_3$- and N_2-$H_2\cdots O_4$ hydrogen bond) and two weaker C-H$\cdots$O interactions (C_{27}-$H_{27}\cdots O_1$ and C_{17}-$H_{17A}\cdots O_5$ hydrogen bond), which generated two types of helices propagated along the b axis (Figure 2c,d).

Table 2. Main hydrogen bonding parameters of two multicomponent crystals of ARI.

	D-H⋯A	D-H/Å	H⋯A/Å	D⋯A/Å	D-H⋯A/°
ARI-SAL Salt	$C_{16}\text{-}H_{16A}\cdots Cl_1$	0.97	2.675	3.248	118
	$O_5\text{-}H_5\cdots O_4$	0.82	1.789	2.497	143
	$N_2\text{-}H_2\cdots O_3$	0.98	1.691	2.670	177
	$N_2\text{-}H_2\cdots O_4$	0.98	2.503	3.122	121
	$N_1\text{-}H_1\cdots O_1^{\text{i}}$	0.86	2.062	2.908	168
	$C_{17}\text{-}H_{17A}\cdots O_5^{\text{ii}}$	0.97	2.542	3.467	160
	$C_{27}\text{-}H_{27}\cdots O_1^{\text{iii}}$	0.93	2.360	3.287	174
	Symmetry transformation: ^{i}: $-x+2, -y, -z+1$; ^{ii}: $-x+1, y-1/2, -z+3/2$; ^{iii}: $-x+2, y+1/2, -z+3/2$.				
ARI-ADI salt acetone hemisolvate	$C_{16}\text{-}H_{16A}\cdots Cl_1$	0.97	2.595	3.214	121
	$N_1\text{-}H_1\cdots O_1^{\text{i}}$	0.86	2.023	2.869	167
	$N_2\text{-}H_2\cdots O_3$	0.98	1.729	2.685	164
	$C_{15}\text{-}H_{15A}\cdots Cl_1^{\text{ii}}$	0.97	2.957	3.448	112
	$C_{11}\text{-}H_{11A}\cdots \pi(Cg)^{\text{iii}}$	0.97	2.96	3.843	151
	Symmetry transformation ^{i}: $-x+1, -y-1, -z$; ^{ii}: $x+1, y, z$; ^{ii}: $x, -1+y, z$. Cg is the centroid of $C_{18}/C_{19}/C_{20}/C_{21}/C_{22}/C_{23}$ atoms				

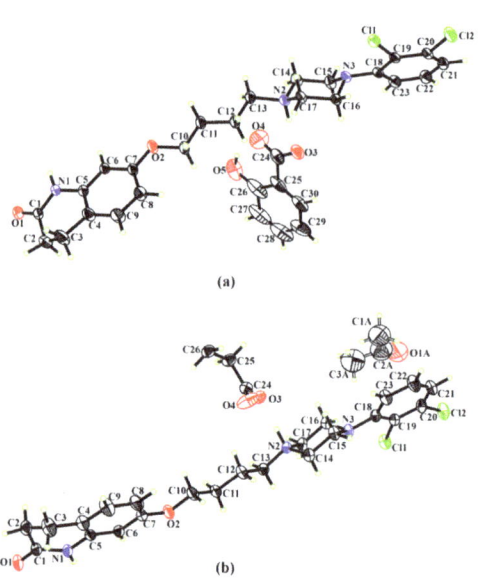

Figure 1. The molecular structures of two ARI multicomponent crystals with displacement ellipsoids drawn at the 30% probability level: (**a**) ARI-SAL salt; (**b**) ARI-ADI salt acetone hemisolvate.

ARI-ADI salt acetone hemisolvate. The ARI-ADI salt acetone hemisolvate crystallized in the triclinic P-1 space group. The ORTEP diagram contained an aripiprazole cation, half of an adipic acid counter-ion (bisected by an inversion center), and half of an acetone molecule (Figure 1b). The solvent acetone molecule was disordered with a position occupancy of 0.5 due to the location on the symmetric center, and acted as a filler in spatial network. The centrosymmetric $N_1\text{-}H_1\cdots O_1$ dimer $R_2^2(8)$ motif with its inversion-related molecule also can be found and formed a 1D ribbon by the strongest charged $N_2^+\text{-}H_2\cdots O_3$ hydrogen bond (Figure 3a). The 1D ribbon interacted with its adjacent ribbon and then formed a 2D sheet through a $C_{15}\text{-}H_{15A}\cdots Cl_1$ hydrogen bond (Figure 3b). Finally, the 2D sheets were arranged into the three-dimensional network by a weak $C_{11}\text{-}H_{11A}\cdots \pi$ hydrogen bond (Figure 3c).

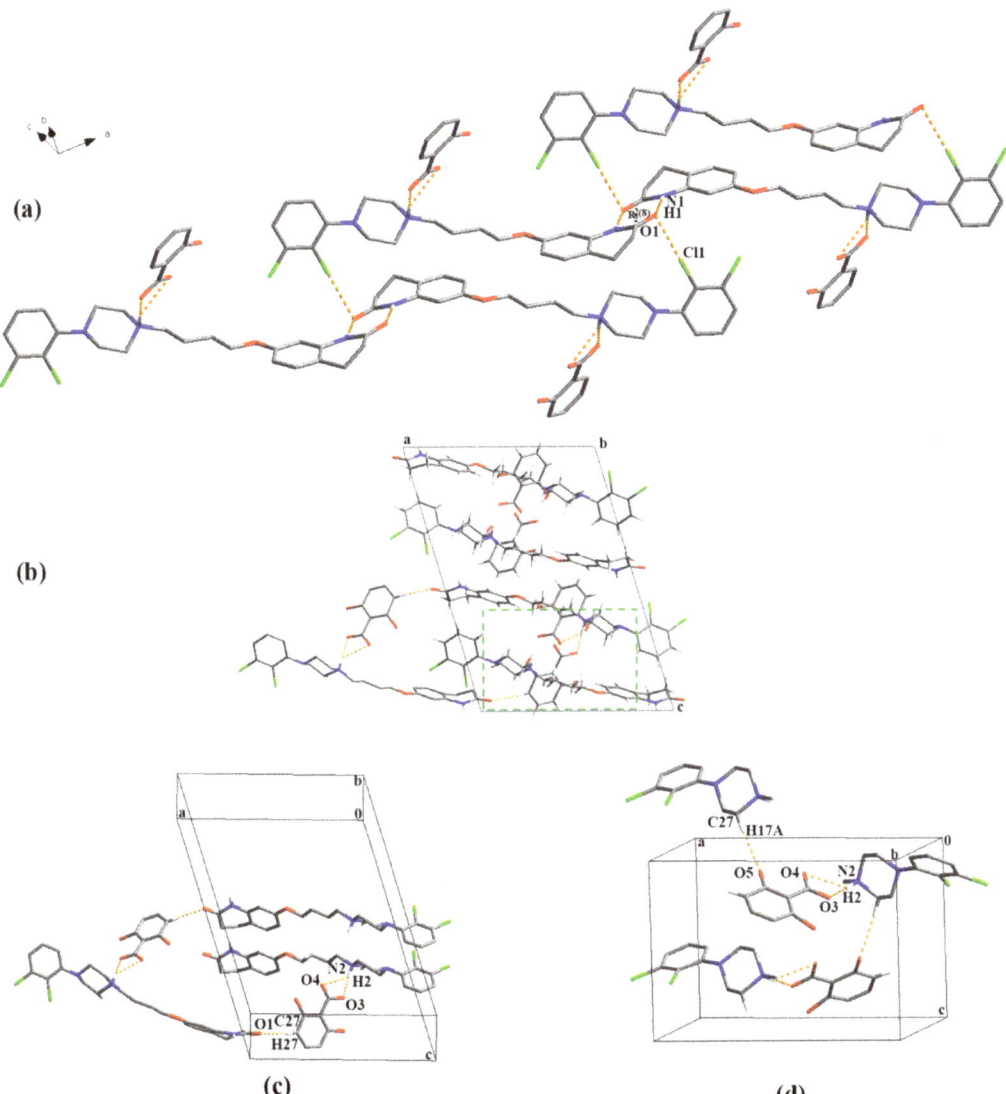

Figure 2. Packing diagrams of ARI-SAL salt: (**a**) the 1D ribbon was formed by N_1-H_1···O_1 hydrogen bond and short Cl_1···O_1 interaction; (**b**) packing diagram for ARI-SAL salt viewed down the b axis and the green dashed wireframe indicates two different types of helices; (**c**) the type-1 helix was propagated by the bifurcated N-H···O hydrogen bond and weaker C_{27}-H_{27}···O_1 interactions; (**d**) the type-2 helix was propagated by the bifurcated N-H···O hydrogen bond and C_{17}-H_{17A}···O_5 interactions (only a partial aripiprazole molecule was drawn for clarity).

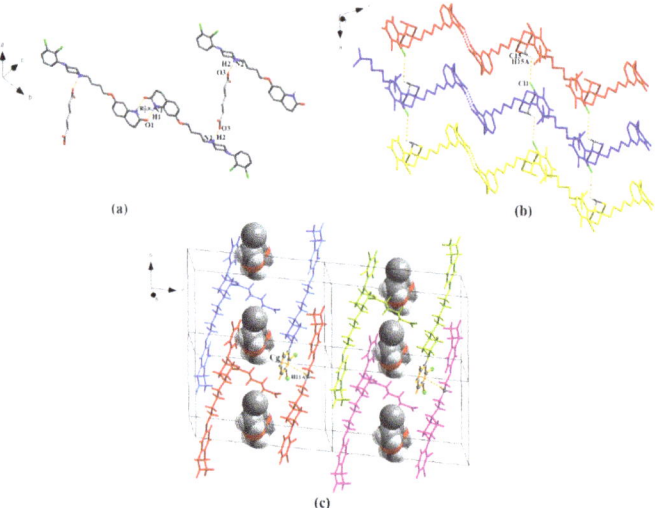

Figure 3. Packing diagrams of ARI-ADI salt acetone hemisolvate: (**a**) the 1D ribbon formed by N_1-H_1···O_1 and N_2-H_2···O_3 hydrogen bond; (**b**) the 2D sheet formed by C_{15}-H_{15A}···Cl_1 hydrogen bond; (**c**) packing diagram for ARI-ADI salt acetone hemisolvate formed by C_{11}-H_{11A}···Cg hydrogen bond.

3.1.2. PXRD Analysis

The ARI multicomponent crystals were synthesized and characterized by PXRD. As seen in Figure 4, their experimental patterns show significant differences compared with those of ARI, which are confirmed as new crystalline forms due to the absence of the characteristic peak of ARI at 20.483° and 22.199°. The characteristic PXRD peaks of the ARI-SAL salt were 12.948°, 18.401°, and 19.849°, while those of the ARI-ADI salt were 25.079°, 14.976°, and 18.256°, which were different from those of its acetone hemisolvte with characteristic PXRD peaks at 18.197° and 24.897°. Importantly, the experimental patterns of ARI-ADI salt acetone hemisolvate and ARI-SAL salt coincided well with their own calculated PXRD patterns from single crystal structures, indicating that the bulk pure samples were prepared successfully.

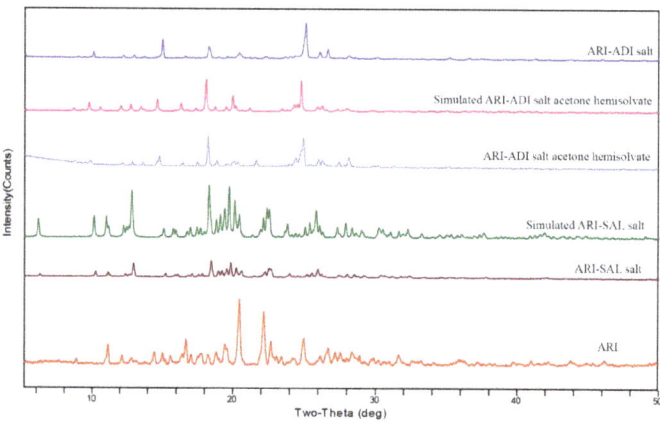

Figure 4. PXRD patterns of ARI and its crystal salts.

3.1.3. DSC Analysis

As reported by Zhao Yanxiao [18], the DSC thermogram (Figure 5) of ARI exhibited two endothermic peaks and a small exothermic peak. The endothermic peak at 138.94 °C corresponded to the melting point of form III, while the second endothermic peak at 147.79 °C represented the melting point of form I, demonstrating that form III might be transformed to form I during the heating process. As expected, the DSC thermograms of ARI multicomponent crystals were distinct from those of ARI and CCF. A single melting endothermic peak was observed at 182.38 °C for the ARI-SAL salt, indicating the formation of pure solid phase. For the ARI-ADI salt acetone hemisolvate, a weak endothermic event of desolvation and a sharp melting endothermic peak were observed at 57.43 °C and 121.18 °C, respectively, which implied that it had a poor thermodynamic stability and may transform into a desolvated salt. As we intended, the ARI-ADI salt showed a single melting endothermic peak at 120.36 °C.

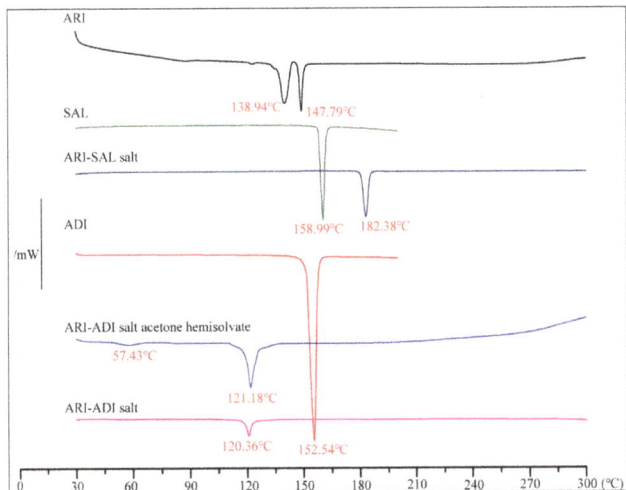

Figure 5. DSC thermograms of ARI, CCF, and its crystal salts.

3.2. Theoretical Calculation

3.2.1. Acid Dissociation Constant (pKa)

ARI is a weak base with a pKa of 7.46, and both salicylic acid and adipic acid are weak acid excipients with pKa values of 2.79 and 3.92, respectively. According to the ΔpKa rule, if the difference of pKa between active pharmaceutical ingredients and CCF was greater than 3, the binary mixture tended to form a salt type [25]. Therefore, we predicted that both of the synthesized multicomponent crystals of ARI were salts, which could be confirmed by especially single-crystal X-ray diffraction.

3.2.2. Molecular Electrostatic Potential Surface

MEPs can provide some clear information in terms of the electrophilic and nucleophilic attack region of the molecule. The values of MEP, related to the strength of intermolecular interactions, can predict the formation possibility of multicomponent crystals. The MEPs mapped on van der waals (VDW) surface of ARI, SAL, and ADI are given in Figure 6, where the red color represents a positive electrical potential and the blue color represents a negative electrical potential. The pKa values were calculated by MarvinSketch and are presented in the black color.

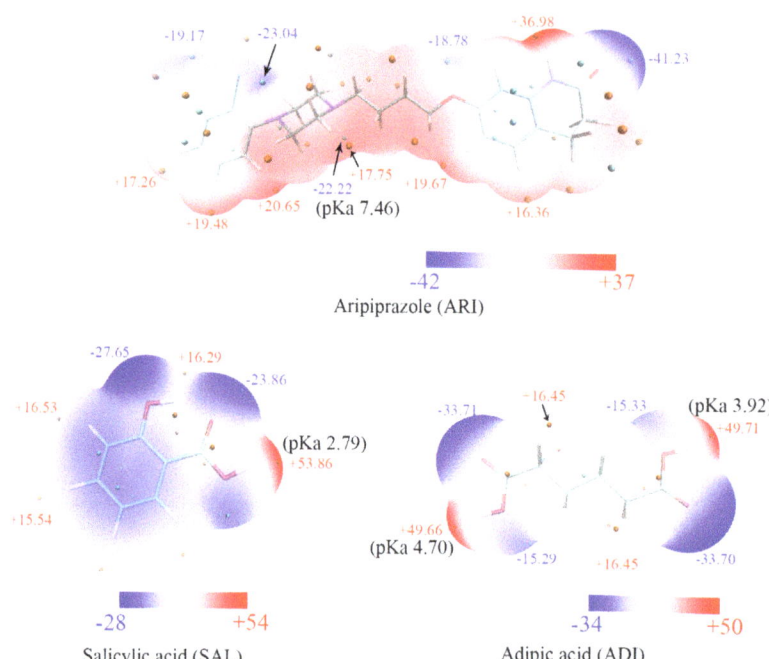

Figure 6. Electrostatic potential mapped onto the 0.001 a.u. electron density isosurface of ARI, SAL, and ADI. The maximal and minimal extreme points of electrical potential are marked with red and blue, repectively. The pKa values are presented in black.

As reported by Zhao Yanxiao [18], the global maxima and minima values of the electrostatic potential on the surface in ARI were +36.98 and −41.23 kcal/mol, which corresponded to carbonyl and amino group of the lactam ring, respectively, and formed the N-H···O homodimers as evidenced in this study. The N_2 and N_3 atoms of the piperazinyl group had little difference in electrical potential, but the N_3 atom had the strong p-π conjugative effect with adjacent benzene ring. As a result, the N_2 atom was vulnerable to electrophilic attack, and no ARI variant containing an exclusive protonation of the N_3 atom was observed [16]. The global maxima value of the electrostatic potential on the surface in SAL was +53.86 kcal/mol, which corresponded to carboxylate group with a low pKa value (pKa = 2.79). As result, the deprotonated N^+-H···O hydrogen bond was formed in ARI-SAL salt. For ADI, the global and secondary maxima values were +49.71 and +49.66 kcal/mol, which corresponded to both of the carboxylate groups with identical nucleophilic reactivity, respectively. As expected, the deprotonated N^+-H···O hydrogen-bonding contacts occurred in real ARI-ADI salt, where the stoichiometric ratio of ARI to ADI was 1:0.5.

3.2.3. Hirshfeld Surface Analysis

The 3D Hirshfeld surface and 2D fingerprint plots of two multicomponent crystals of ARI are given in Figure 7. In their 3D Hirshfeld surface, the large and deep red spots correspond to the close-contact N-H···O interactions. The H···H contacts appear as scattered points in the middle region of the 2D fingerprint plots, which were the predominate type of interactions. The relative percentage contributions of H···H contacts in theARI-SAL salt were markedly lower than that of the ARI-ADI salt acetone hemisolvate and compensated by the C···H/H···C contacts shown as a pair of "wings". However, these interactions displayed higher interatomic distance. Interactions with less interatomic distance are presented in fingerprint plots as pointy regions. The reciprocal O···H contacts, corresponding to O-H···O, N-H···O, and C-H···O interactions, are presented as long, sharp, symmetrical spikes. Additionally, the Cl···H contacts, corresponding to the Cl···H-C, Cl···H-O and Cl···H-N interactions, appear as a pair of "wings" and have equivalent relative percentage contributions to their own Hirshfeld surfaces.

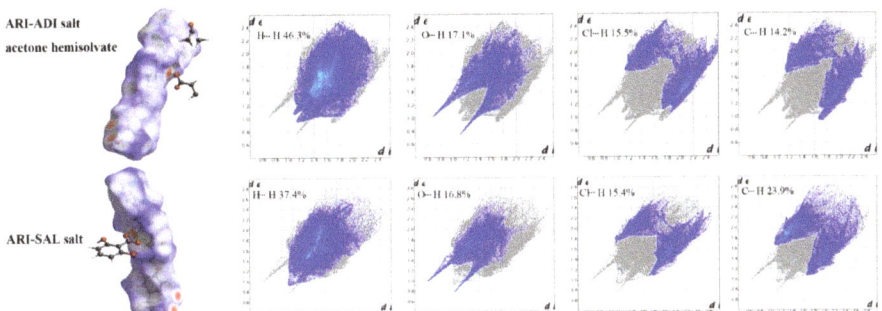

Figure 7. The 3D Hirshfeld surface and 2D fingerprint plots of two multicomponent crystals of ARI.

3.3. Powder Hygroscopicity

The moisture absorption is a crucial parameter that should be considered during the process of drug development, production, and storage. Therefore, the powder hygroscopicity was measured under 25 °C/80% RH condition. Ten days later, the hygroscopic gain of ARI was 0.32 ± 1.30%. Meanwhile, ARI-SAL salt and ARI-ADI salt and its acetone hemisolvate showed greater hygroscopicity, with hygroscopic gains of 2.60 ± 0.78%, 1.36 ± 0.32%, and 1.13 ± 0.7%, respectively. This might be presumably due to the presence of a polar group (i.e., –OH and –COOH) introduced by synthesizing carboxylate salts, leading to the formation of hydrogen bonds with water molecules [33].

3.4. Stability Test

Moisture and high temperatures may cause the transformation of drug crystal forms and degradation, especially for solvates. Thus, close attention was paid to their stability under high humidity and high temperatures in this work. Figure 8 illustrates that the ARI-SAL salt was stable, while the ARI-ADI salt acetone hemisolvate was unstable under a high temperature. There were significant differences at the two-theta range of 19° to 27° compared with the starting PXRD pattern. In conjunction with DSC patterns of ARI-ADI salt acetone hemisolvate, we speculated that desolvation occurred under high temperatures. Unexpectedly, the PXRD pattern of the ARI-ADI salt was inconsistent with that of its acetone hemisolvate under high temperatures, demonstrating that ARI-ADI salt might be unstable under high temperatures.

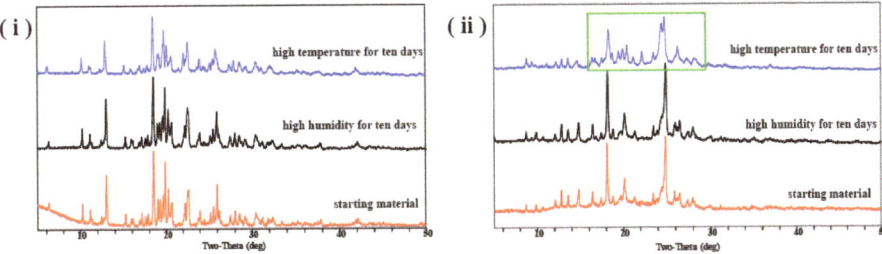

Figure 8. Results of stability testing of ARI-SAL salt (i) and ARI-ADI salt acetone hemisolvate (ii).

3.5. Solubility

3.5.1. Equilibrium Solubility Test

The results of the equilibrium solubility experiments are summarized in Figure 9. The equilibrium solubility of the ARI-ADI salt in water, hydrochloric acid buffer, and acetate acid-sodium acetate buffered solution were 0.13, 0.63, and 0.15 mg/mL, respectively, which was much better than that of the ARI and ARI-SAL salt but slightly smaller than that of the ARI-ADI salt acetone hemisolvate. However, the equilibrium solubility of the ARI-SAL salt was even worse than that of the ARI in both the hydrochloric acid buffer solution and acetate acid-sodium acetate buffered solution, though slightly better in water. In addition, the equilibrium solubility of ARI and its salts in the phosphate buffer solution were too low to be acquired accurately.

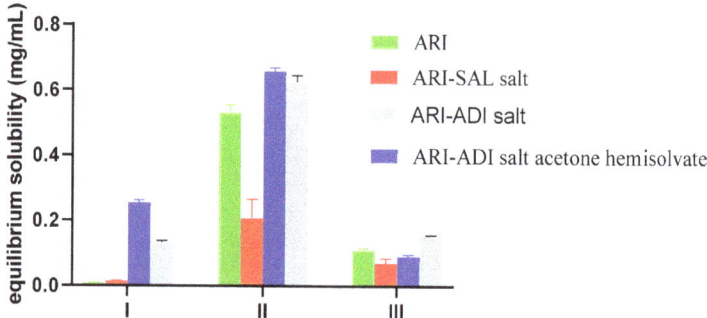

Figure 9. The equilibrium solubilities of ARI and its crystal salts in different solutions. **I**: water (pH 7.0); **II**: hydrochloric acid buffer (pH 1.2); **III**: acetate acid-sodium acetate buffered solution (pH 4.5).

3.5.2. IDR Test

Given its kinetic nature, IDR assumes a better correlation with in vivo drug dissolution dynamics than solubility [34]. Thus, IDR studies were carried out at pH 1.2 in this work, and the results are summarized in Figure 10. The ARI held an IDR of 0.5051 mg·cm^{-2}·min^{-1}, while the IDR of the ARI-ADI salt, slightly lower than that of its acetone hemisolvate, significantly increased to 0.9263 mg·cm^{-2}·min^{-1} and was about twice as much as that of the ARI. Not unexpectedly, the IDR of the ARI-SAL salt was only 0.3745 mg·cm^{-2}·min^{-1}, significantly lower than that of the ARI.

These findings indicated that ARI-ADI salt and its acetone hemisolvate had predictable advantages in vivo absorption over ARI-SAL salt and the untreated ARI. This may be due to many causes. First, the relatively higher hygroscopicity and lower melting point may lead to easier dissolution in an aqueous solvent, because of it demanding less lattice energy to break. Molecular constituents were also one of the important causes. The CCF of ARI-ADI salt was more soluble than that of the ARI-SAL salt, and their solubility behaviors follow the rule of thumb that the greater solubility of the CCF, the more soluble the salt will be [18]. Finally, their spatial structures may also be associated with their differences in solubility. The ADI molecules linked with the solvent molecules to form a hydrophilic layer. As a result, water molecules permeate more easily into the layer stacking spatial structure and cause it to disintegrate in aqueous solvent.

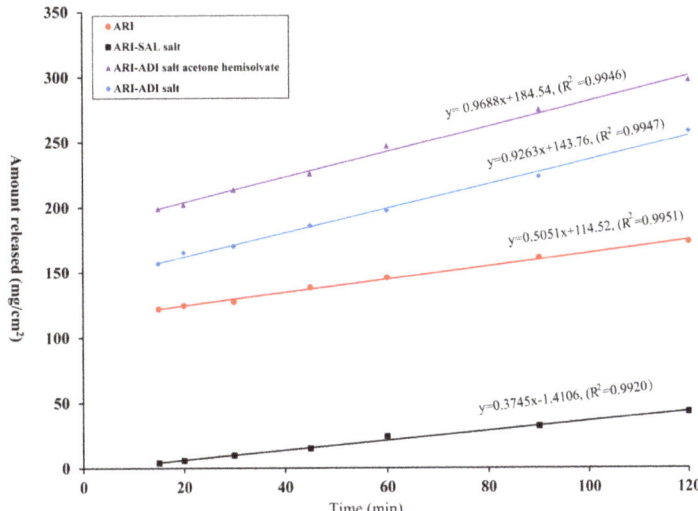

Figure 10. Intrinsic dissolution profiles of the ARI and its crystal salts in hydrochloric acid buffer (pH 1.2).

4. Conclusions

In this study, a new ARI-ADI salt and its acetone hemisolvate were successfully synthesized and fully characterized by various methods, along with a reported ARI-SAL salt. Their crystal structures revealed that the proton transferring occurred from a carboxylic group of CCF to the N_2 atoms of the piperazine region from the ARI. Structural analysis showed that the ARI-SAL salt was arranged helically by two helices propagated along the b axis, while the ARI-ADI salt acetone hemisolvate presented a typical layer stacking with the help of weak C-H···Cl and C-H···π hydrogen bonds. The reactivity of ARI and CCF was predicted by MEPS and matched well with their hydrogen bond schemes. Hirshfeld surface analysis was also used to clarify the intermolecular interactions. Furthermore, the ARI-ADI salt had a significant advantage over the ARI and ARI-SAL salt in terms of equilibrium solubility and IDR. This study provided a valuable insight into the formation of multicomponent pharmaceutical salts and presented potential alternative formulations of ARI.

Author Contributions: Conceptualization, D.Y. and J.T.; tests and data interpretation, Q.Z., Z.T., Y.W. and Y.Z.; writing—original draft preparation, Q.Z.; review and editing, Q.Z., Y.L. and G.G. All authors have read and agreed to the published version of the manuscript.

Funding: This research received no external funding.

Data Availability Statement: The data used to support the findings of this study are included within the article.

Conflicts of Interest: The authors declare no conflict of interest.

Appendix A

The supplementary crystallographic data for this paper have been deposited in the Cambridge Crystallographic Data Center and have reference numbers CCDC 1991810 and 2023732. These data can be obtained free of charge via http://www.ccdc.cam.ac.uk (accessed on 26 March 2021).

References

1. Sethiya, A.; Agarwal, D.K.; Agarwal, S. Current Trends in Drug Delivery System of Curcumin and its Therapeutic Applications. *Mini Rev. Med. Chem.* **2020**, *13*, 1190–1232. [CrossRef]
2. Berry, D.J.; Steed, J.W. Pharmaceutical cocrystals, salts and multicomponent systems; intermolecular interactions and property based design. *Adv. Drug Deliv. Rev.* **2017**, *117*, 3–24. [CrossRef] [PubMed]
3. Diniz, L.F.; Carvalho, P.S., Jr.; Pena, S.A.C.; Gonçalves, J.E.; Souza, M.A.C.; de Souza Filho, J.D.; Bomfim Filho, L.F.O.; Franco, C.H.J.; Diniz, R.; Fernandes, C. Enhancing the solubility and permeability of the diuretic drug furosemide via multicomponent crystal forms. *Int. J. Pharm.* **2020**. [CrossRef] [PubMed]
4. Clements, M.; Roex, T.L.; Blackie, M. Multicomponent Crystal Systems of Known Antimalarial Drug Molecules. *ChemMedChem* **2015**, *10*, 1786–1792. [CrossRef]
5. Sanphui, P.; Mishra, M.K.; Ramamurty, U.; Desiraju, G.R. Tuning mechanical properties of pharmaceutical crystals with multicomponent crystals: Voriconazole as a case study. *Mol. Pharm.* **2015**, *12*, 889–897. [CrossRef] [PubMed]
6. Elder, D.P.; Holm, R.; Diego, H.L. Use of pharmaceutical salts and cocrystals to address the issue of poor solubility. *Int. J. Pharm.* **2013**, *453*, 88–100. [CrossRef] [PubMed]
7. Prommer, E. Aripiprazole. *Am. J. Hosp. Palliat. Care* **2017**, *34*, 180–185. [CrossRef]
8. Cuomo, A.; Beccarini Crescenzi, B.; Goracci, A.; Bolognesi, S.; Giordano, N.; Rossi, R.; Facchi, E.; Neal, S.M.; Fagiolini, A. Drug safety evaluation of aripiprazole in bipolar disorder. *Expert Opin. Drug Saf.* **2019**, *18*, 455–463. [CrossRef]
9. Borrego-Sánchez, A.; Sánchez-Espejo, R.; Albertini, B.; Passerini, N.; Cerezo, P.; Viseras, C.; Sainz-Díaz, C.I. Ground Calcium Carbonate as a Low Cost and Biosafety Excipient for Solubility and Dissolution Improvement of Praziquantel. *Pharmaceutics* **2019**, *11*, 533. [CrossRef]
10. Lyszczarz, E.; Hofmanova, J.; Szafraniec-Szczesny, J.; Jachowicz, R. Orodispersible films containing ball milled aripiprazole-poloxamer(R)407 solid dispersions. *Int. J. Pharm.* **2020**. [CrossRef]
11. Gao, L.; Zhang, X.R.; Chen, Y.F.; Liao, Z.L. A new febuxostat imidazolium salt hydrate: Synthesis, crystal structure, solubility, and dissolution study. *J. Mol. Struct.* **2019**, *1176*, 633–640. [CrossRef]
12. Miraglia, N.; Agostinetto, M.; Bianchi, D.; Valoti, E. Enhanced oral bioavailability of a novel folate salt: Comparison with folic acid and a calcium folate salt in a pharmacokinetic study in rats. *Minerva Ginecol.* **2016**, *68*, 99–105.
13. Freire, E.; Polla, G.; Baggio, R. Aripiprazole salts I Aripiprazole nitrate. *Acta Crystallogr. C* **2012**, *68 Pt 4*, o170–o173. [CrossRef]
14. Freire, E.; Polla, G.; Baggio, R. Aripiprazole salts. II. Aripiprazole perchlorate. *Acta Crystallogr. C* **2012**, *68 Pt 6*, o235–o239. [CrossRef]
15. Freire, E.; Polla, G.; Baggio, R. Aripiprazole salts. III. Bis(aripiprazolium) oxalate-oxalic acid (1/1). *Acta Crystallogr. C* **2013**, *69 Pt 2*, 186–190. [CrossRef]
16. Freire, E.; Polla, G.; Baggio, R. Aripiprazole salts IV. Anionic plus solvato networks defining molecular conformation. *J. Mol. Struct.* **2014**, *1068*, 43–52. [CrossRef]
17. Nanubolu, J.B.; Sridhar, B.; Ravikumar, K.; Cherukuvada, S. Adaptability of aripiprazole towards forming isostructural hydrogen bonding networks in multi-component salts: A rare case of strong O–H···O– ↔ weak C–H···O mimicry. *CrystEngComm* **2013**, *15*, 4321–4335. [CrossRef]
18. Zhao, Y.; Sun, B.; Jia, L.; Wang, Y.; Wang, M.; Yang, H.; Qiao, Y.; Gong, J.; Tang, W. Tuning Physicochemical Properties of Antipsychotic Drug Aripiprazole with Multicomponent Crystal Strategy Based on Structure and Property Relationship. *Cryst. Growth. Des.* **2020**, *20*, 3747–3761. [CrossRef]
19. SAINT Version 7.68A. *Software for the CCD Detector System*; Bruker AXS Inc: Madison, WI, USA, 2009.
20. Sheldrick, G.M. *SADABS, Bruker/Siemens Area Detector Absorption Correction Program*; University of Göttingen: Göttingen, Germany, 1998.
21. Sheldrick, G.M. A short history of SHELX. *Acta Crystallogr. A* **2008**, *64*, 112–122. [CrossRef] [PubMed]
22. Sheldrick, G.M. *SHELXL2015, Program for Crystal Structure Refinement*; University of Göttingen: Göttingen, Germany, 2015.
23. Pennington, W. DIAMOND—Visual Crystal Structure Information System. *J. Appl. Crystallogr.* **1999**, *32*, 1028–1029. [CrossRef]
24. Sedghiniya, S.; Soleimannejad, J.; Janczak, J. The salt-cocrystal continuum in salicylic acid-adenine: The influence of crystal structure on proton-transfer balance. *Acta Crystallogr. C Struct. Chem.* **2019**, *75 Pt 4*, 412–421. [CrossRef]
25. Sanphui, P.; Bolla, G.; Nangia, A. High Solubility Piperazine Salts of the Nonsteroidal Anti-Inflammatory Drug (NSAID) Meclofenamic Acid. *Cryst. Growth. Des.* **2012**, *12*, 2023–2036. [CrossRef]
26. Khan, M.F.; Nahar, N.; Rashid, R.B.; Chowdhury, A.; Rashid, M.A. Computational investigations of physicochemical, pharmacokinetic, toxicological properties and molecular docking of betulinic acid, a constituent of Corypha taliera (Roxb.) with Phospholipase A2 (PLA2). BMC complement. *Altern. Med.* **2018**, *18*, 48–62. [CrossRef]
27. Lu, T.; Chen, F. Multiwfn: A multifunctional wavefunction analyzer. *J. Comput. Chem.* **2012**, *33*, 580–592. [CrossRef] [PubMed]
28. Manzetti, S.; Lu, T. The geometry and electronic structure of Aristolochic acid: Possible implications for a frozen resonance. *J. Phys. Org. Chem.* **2013**, *26*, 473–483. [CrossRef]
29. Zhang, X.; Zhou, L.; Wang, C.; Li, Y.; Wu, Y.; Zhang, M.; Yin, Q. Insight into the Role of Hydrogen Bonding in the Molecular Self-Assembly Process of Sulfamethazine Solvates. *Cryst. Growth. Des.* **2017**, *17*, 6151–6157. [CrossRef]
30. Spackman, M.A.; Jayatilaka, D. Hirshfeld Surface Analysis. *CrystEngComm* **2009**, *11*, 19–32. [CrossRef]

31. Yang, D.; Cao, J.; Jiao, L.; Yang, S.; Zhang, L.; Lu, Y.; Du, G. Solubility and stability advantages of a new cocrystal of berberine chloride with fumaric acid. *ACS Omega* **2020**, *4*, 8283–8292. [CrossRef] [PubMed]
32. Steiner, T. Competition of hydrogen-bond acceptors for the strong carboxyl donor. *Acta Crystallogr. B* **2001**, *57*, 103–106. [CrossRef] [PubMed]
33. Newman, A.W.; Reutzel-Edens, S.M.; Zografi, G. Characterization of the "hygroscopic" properties of active pharmaceutical ingredients. *J. Pharm. Sci.* **2008**, *97*, 1047–1059. [CrossRef] [PubMed]
34. Shevchenko, A.; Bimbo, L.M.; Miroshnyk, I.; Haarala, J.; Jelínková, K.; Syrjänen, K.; Veen, B.; Kiesvaara, J.; Santos, H.A.; Yliruusi, J. A new cocrystal and salts of itraconazole: Comparison of solid-state properties stability and dissolution behavior. *Int. J. Pharm.* **2012**, *436*, 403–409. [CrossRef] [PubMed]

Article

Furosemide/Non-Steroidal Anti-Inflammatory Drug–Drug Pharmaceutical Solids: Novel Opportunities in Drug Formulation

Francisco Javier Acebedo-Martínez [1], Carolina Alarcón-Payer [2], Lucía Rodríguez-Domingo [1,3], Alicia Domínguez-Martín [3], Jaime Gómez-Morales [1] and Duane Choquesillo-Lazarte [1,*]

[1] Laboratorio de Estudios Cristalográficos, IACT, CSIC-Universidad de Granada, Avda. de las Palmeras 4, 18100 Armilla, Spain; j.acebedo@csic.es (F.J.A.-M.); luciard13@hotmail.com (L.R.-D.); jaime@lec.csic.es (J.G.-M.)
[2] Servicio de Farmacia, Hospital Universitario Virgen de las Nieves, 18014 Granada, Spain; carolina.alarconpayer@gmail.com
[3] Department of Inorganic Chemistry, Faculty of Pharmacy, University of Granada, 18071 Granada, Spain; adominguez@ugr.es
* Correspondence: duane.choquesillo@csic.es

Abstract: The design of drug–drug multicomponent pharmaceutical solids is one the latest drug development approaches in the pharmaceutical industry. Its purpose is to modulate the physicochemical properties of active pharmaceutical ingredients (APIs), most of them already existing in the market, achieving improved bioavailability properties, especially on oral administration drugs. In this work, our efforts are focused on the mechanochemical synthesis and thorough solid-state characterization of two drug–drug cocrystals involving furosemide and two different non-steroidal anti-inflammatory drugs (NSAIDs) commonly prescribed together: ethenzamide and piroxicam. Besides powder and single crystal X-ray diffraction, infrared spectroscopy and thermal analysis, stability, and solubility tests were performed on the new solid materials. The aim of this work was evaluating the physicochemical properties of such APIs in the new formulation, which revealed a solubility improvement regarding the NSAIDs but not in furosemide. Further studies need to be carried out to evaluate the drug–drug interaction in the novel multicomponent solids, looking for potential novel therapeutic alternatives.

Keywords: drug–drug cocrystal; furosemide; ethenzamide; piroxicam; mechanochemical synthesis

Citation: Acebedo-Martínez, F.J.; Alarcón-Payer, C.; Rodríguez-Domingo, L.; Domínguez-Martín, A.; Gómez-Morales, J.; Choquesillo-Lazarte, D. Furosemide/Non-Steroidal Anti-Inflammatory Drug–Drug Pharmaceutical Solids: Novel Opportunities in Drug Formulation. *Crystals* **2021**, *11*, 1339. https://doi.org/10.3390/cryst11111339

Academic Editor: Aidar T. Gubaidullin

Received: 18 October 2021
Accepted: 31 October 2021
Published: 2 November 2021

Publisher's Note: MDPI stays neutral with regard to jurisdictional claims in published maps and institutional affiliations.

Copyright: © 2021 by the authors. Licensee MDPI, Basel, Switzerland. This article is an open access article distributed under the terms and conditions of the Creative Commons Attribution (CC BY) license (https://creativecommons.org/licenses/by/4.0/).

1. Introduction

Diuretic drugs aim to regulate the volume and composition of body fluids by increasing the rate of urine flow and sodium excretion. They are widely used in clinics for the treatment of edematous disorders, such as those associated with congestive heart failure, as well as liver or renal failure and hypertension [1,2]. Furosemide (FUR, Scheme 1), 4-chloro-2-[(2-furanylmethyl)-amino]-5-sulfamoylbenzoic acid, is classified as a high ceiling loop diuretic drug. Its mechanism of action is related to the inhibition of the sodium-potassium-2chloride co-transporter (Na^+-K^+-$2Cl^-$) located in the thick ascending limb of the loop of Henle in the renal tubule.

According to the Biopharmaceutics Classification System (BCS), FUR belongs to class IV drug, defined by low solubility and low permeability values [3]. Indeed, furosemide is almost insoluble in water [4], which results in significant intraindividual variations in absorption and very poor oral bioavailability [5]. Despite this relevant drawback, FUR has shown great efficacy, hence it is highly used in therapeutics worldwide, including chronic treatments. Thereby, the development of improved oral formulations of furosemide, which aim to achieve higher bioavailability, are certainly relevant for the pharmaceutical industry.

The design of multicomponent pharmaceutical solids is actually one of the latest research strategies in the development of new drug alternatives in the pharmaceutical industry [6]. They can be defined as crystalline materials in which at least one component is an active pharmaceutical ingredient (API). The other components, incorporated in the crystal lattice—so-called coformers—must be found in a stoichiometric ratio and considered pharmaceutically acceptable—i.e., included in the Generally Recognized as Safe (GRAS) list within the FDA's "Substances Added to Food" Inventory. APIs and coformers recognize themselves by different kind of non-covalent intermolecular interactions, so-called supramolecular synthons, mainly H-bonds, which organization has a profound impact on the intimate 3D structure of the solid and therefore on its macroscopic physicochemical properties. The development of this novel strategy is rather interesting because allows industry to save money compared to the traditional drug development scheme, still guarantying the possibility of generating intellectual property rights [7]. In this context, there has been reported several studies devoted to studying pharmaceutical cocrystals and salts of furosemide [8–13].

Scheme 1. Chemical formula of furosemide (FUR), ethenzamide (ETZ), and piroxicam (PRX).

One of the most recent approaches in the development of multicomponent pharmaceutical solids is the concurrent administration of two or more APIs, leading to drug–drug or co-drug pharmaceutical solids [14]. APIs within the formulation might have similar or different mechanisms of actions, but always looking for a synergic effect, either targeting one metabolic pathway at different levels or different pathways related to a particular disease.

Along with diuretics, nonsteroidal anti-inflammatory drugs (NSAIDs) are also widely prescribed worldwide. Interestingly, the combination of diuretics—particularly furosemide—and NSAIDs is rather common, especially among the elderly. However, although not contraindicated, there is clinical evidence on the moderate interaction between these two kinds of drugs. The use of NSAIDs may decrease natriuretic response to loop diuretics, thus reducing their efficacy and resulting in adverse effects on patients with different edematous states. In addition, some NSAIDs may also show adverse nephrotoxic effects, which may be exacerbated by diuretic therapy [15–17]. In these cases, dose adjustments or special monitoring of the renal function and blood pressure are required for safety's sake. Unfortunately, the insights of such interactions are still poorly understood because they do not seem to follow the same mechanism for all combination of drugs, some of them being associated with the suppression of plasma renin activity or impaired synthesis of vasodilator prostaglandins.

Since the concurrent prescription of FUR and NSAIDs is quite common, it is worthwhile exploring the formulation of drug–drug multicomponent pharmaceutical solids involving such a combination, seeking new therapeutical alternatives that would improve the bioavailability of the APIs and/or reduce the abovementioned drug–drug interactions. In this work, the synthesis and physicochemical characterization of two different drug–drug pharmaceutical solids, including the loop diuretic furosemide and one NSAID drug: ethenzamide (ETZ, 2-ethoxybenzamide) or piroxicam (PRX, 4-hydroxyl-2- methyl-N-2-pyridinyl-2H-1,2,-benzothiazine-3-carboxamide 1,1-dioxide) are reported (Scheme 1). To the best of our knowledge, there are no conclusive studies on the interaction between furosemide and ethenzamide, while one study was reported on piroxicam–furosemide drug interaction in the late 1980s [18].

2. Materials and Methods

2.1. Materials

Furosemide, ethenzamide, piroxicam, and solvents used are commercially available from Sigma-Aldrich. All solvents were used as received without additional purification.

2.2. Coformer Selection

A search of the Cambridge Structural Database (CSD) [19] was conducted to identify the coformers with complementary functional groups that can serve as components for molecular recognition with FUR.

The excess enthalpy (H_{ex}) of mixing between FUR and selected coformers was calculated using COSMOquick software [20] (COSMOlogic, Germany, Version 1.4).

2.3. General Procedure for Mechanochemical Synthesis

Mechanochemical syntheses of cocrystals were conducted by liquid-assisted grinding (LAG) in a Retsch MM200 ball mill (Retsch, Haan, Germany) operating at 25 Hz frequency using stainless steel jars along with stainless steel balls of 7 mm diameter. All syntheses were repeated to ensure reproducibility. For liquid-assisted grinding screening, methanol was used as solvent.

Synthesis of FUR–ETZ: a mixture of FUR (165.37 mg, 0.50 mmol) and ETZ (82.59 mg, 0.50 mmol) in a 1:1 stoichiometric ratio was placed in a 10 mL stainless steel jar along with 150.0 µL of methanol and two stainless steel balls of 7 mm diameter. The mixture was then milled for 30 min.

Synthesis of FUR–PRX: a mixture of FUR (165.37 mg, 0.50 mmol) and PRX (165.67 mg, 0.50 mmol) in a 1:1 stoichiometric ratio was placed in a 10 mL stainless steel jar along with 150.0 µL of methanol and two stainless steel balls 7 mm in diameter. The mixture was then milled for 30 min.

2.4. Powder X-ray Diffraction (PXRD)

Powder X-ray diffraction data were collected using a Bruker D8 Advance Vario diffractometer (Bruker-AXS, Karlsruhe, Germany) equipped with a LYNXEYE detector and Cu-Kα1 radiation (1.5406 Å). All the profile fittings were conducted using the software Diffrac.TOPAS 6.0 [21]. The bulk phase purity was checked by Le Bail profile fitting, using cell parameters from structural crystallographic information of the constitutive phases—namely FUR, ETZ, and PRX—as well as the new reported phases. In these fittings, only the background, unit cell parameters and zero error were refined. Rwp values obtained in all cases demonstrate an excellent agreement between the structural model and the bulk phase measured by powder diffraction.

2.5. Preparation of Single Crystals

Single crystals were grown by solvent evaporation at room temperature using the polycrystalline material obtained from mechanical synthesis. Suitable crystals for X-ray diffraction studies were obtained from recrystallization in saturated solutions after approximately 2 days: methanol and acetone for FUR–ETZ and ethanol for FUR–PRX.

2.6. Single-Crystal X-ray Diffraction (SCXRD)

Measured crystals were prepared under inert conditions immersed in perfluoropolyether as protecting oil for manipulation. Suitable crystals were mounted on MiTeGen Micromounts™ (MiTeGen, Ithaca, NY, USA), and these samples were used for data collection. Data for FUR–ETZ and FUR–PRX were collected with a Bruker D8 Venture diffractometer (Bruker-AXS, Karlsruhe, Germany) with graphite monochromated MoKα (FUR–ETZ, λ = 0.71073 Å, at 298(2) K) or CuKα radiation (FUR–PRX, λ = 1.54178 Å, at 298(2) K). The data were processed with APEX3 suite [22]. The structures were solved by Intrinsic Phasing using the ShelXT program [23], which revealed the position of all non-hydrogen atoms. These atoms were refined on F^2 by a full-matrix least-squares procedure using anisotropic displacement parameter [24]. All hydrogen atoms were located in difference Fourier maps and included as fixed contributions riding on attached atoms with isotropic thermal displacement parameters 1.2- or 1.5-times those of the respective atom. The OLEX2 software was used as a graphical interface [25]. Intermolecular interactions were calculated using PLATON [26]. Molecular graphics were generated using Mercury [27]. The crystallographic data for the reported structures were deposited with the Cambridge Crystallographic Data Center as supplementary publication No. CCDC 2114160 and 2114161. Additional crystal data are shown in Table 1. Copies of the data can be obtained free of charge at http://www.ccdc.cam.ac.uk/products/csd/request (accessed on 30 October 2021).

Table 1. Crystallographic data and structure refinement details of FUR cocrystals.

Compound Name	FUR–ETZ	FUR–PRX
Formula	$C_{21}H_{22}ClN_3O_7S$	$C_{27}H_{24}ClN_5O_9S_2$
Formula weight	495.92	662.08
Crystal system	Monoclinic	Monoclinic
Space group	$P2_1/c$	$P2_1/n$
$a/Å$	13.1846 (4)	9.0971 (4)
$b/Å$	9.8733 (3)	23.8637 (10)
$c/Å$	17.1518 (6)	13.7806 (6)
$\alpha/°$	90	90
$\beta/°$	95.776 (2)	99.227 (2)
$\gamma/°$	90	90
$V/Å^3$	2221.41 (12)	2952.9 (2)
Z	4	4
$D_c/g\ cm^{-3}$	1.483	1.489
μ/mm^{-1}	0.315	3.010
F(000)	1032	1368
Reflections collected	32,492	41,427
Unique reflections	5104	5172
R_{int}	0.1392	0.0331
Data/restraints/parameters	5104/0/305	5172/0/406
Goodness-of-fit (F^2)	1.002	1.032
R1 ($I > 2\sigma(I)$)	0.0584	0.0379
wR2 ($I > 2\sigma(I)$)	0.1054	0.0959
Packing coefficient	0.69	0.67

2.7. Stability Test

Slurry experiments were conducted using excess powder samples of each phase in 1 mL of water for 24 h at room temperature in a sealed vial containing a magnetic stirrer. The solids in the vials were collected, filtered, and dried at 35 °C for subsequent analysis by PXRD.

Stability of all the new phases was also studied at accelerated storage condition; 200 mg of each solid was taken in watch glasses and the physical stability was evaluated at 40 °C in 75% relative humidity using a Memmert HPP110 climate chamber (Memmert, Schwabach, Germany). The samples were subjected to the above accelerated stability conditions for 3 days and weekly intervals from 1 week to 8 weeks. PXRD was used to monitor the stability of the solid forms.

2.8. Infrared Spectroscopy

Fourier-transform infrared (FTIR) spectroscopic measurements were performed on a Bruker Tensor 27 FTIR instrument (Bruker Corporation, Billerica, MA, USA) equipped with a single-reflection diamond crystal platinum ATR unit and OPUS data collection program. The scanning range was from 4000 to 400 cm^{-1} with a resolution of 4 cm^{-1}.

2.9. Thermal Analysis

Simultaneous thermogravimetric analysis (TGA) and differential scanning calorimetry (DSC) measurements were performed using a Mettler Toledo TGA/DSC1 thermal analyzer (Mettler Toledo, Columbus, OH, USA). Samples (3–5 mg) were placed into sealed aluminum pans and heated in a stream of nitrogen (100 mL min^{-1}) from 25 to 400 °C at a heating rate of 10 °C min^{-1}.

2.10. Solubility Studies

Solubility studies for pure FUR and each new cocrystal were performed using the Crystal16 equipment (Technobis Crystallization Systems, Alkmaar, The Netherlands) in water PBS at pH 7.4. The equipment is comprised of four individually controlled reactors, each with a working volume of 1 mL, allowing the measurement of cloud and clear points based on the turbidity of 16 aliquots of 1 mL of solution in parallel and automatically. Each composition was heated at 0.5 °C/min to 90 °C with a magnetic stirring rate of 700 rpm, held at this temperature for 10 min and then cooled to 20 °C at 0.5 °C/min. The temperature of dissolution for each compound was measured using different amounts of solid, and the solubility data of the pure components were fitted to the Van't Hoff equation [28] using the CrystalClear software (Technobis Crystallization Systems, Alkmaar, The Netherlands).

3. Results and Discussion

3.1. Coformer Selection

Before the experimental trials, we performed virtual cocrystal screening to improve the success rate. A survey on the Cambridge Structural Database (CSD version 5.42, update 2 from May 2021) based on FUR resulted in 70 hits. After excluding datasets corresponding to FUR polymorphs, the remaining dataset corresponded to multi-component crystals (cocrystals, salts, and solvates), 60 hits. Several authors have reported pharmaceutical salts and cocrystals of FUR. A search of this dataset for drug–drug multi-component crystals revealed a total of 11 systems [8,10–13,29–31]. A common structural feature observed is the key role of the carboxylic group in the interaction with the coformer or counterion and the formation of other synthons involving the sulfonamide group that participates in stabilizing the crystal structure packing. Hence, the abundance of heterosynthons observed in the survey involving carboxylic group follows the order: carboxylic-pyridine (54%) > carboxylic-amide (20%) > carboxylic-imidazole (8%) > carboxylate···piperazinium/carboxylate···ammonium/carboxylate···pyridinium (6%). According to the abovementioned, the main prerequisite for the coformer selection was having the above-referred groups and being a drug. From our library of coformers, two molecules fulfil these criteria: ETZ and PRX. COSMOQuick software was used to validate our selection, predicting the tendency of cocrystal formation based on thermodynamics calculations. This tool calculates the excess enthalpy of formation (H_{ex}) between FUR and the corresponding coformer/drug relative to the pure components in a supercooled liquid phase [32]. It requires the simplified molecular input line entry specification or SMILES of a molecule as input data. Table 2 shows COSMOQuick calculations for a list of candidates to form multi-component crystals with FUR. The list includes our two selected drugs and other coformer molecules involved in the formation of cocrystals/salts reported in the survey. Compounds with negative H_{ex} values show an increased probability of forming cocrystals since H_{ex} is a rough approximation of the free energy of cocrystal formation $\Delta G_{cocrystal}$. The results obtained by COSMOQuick confirm FUR preference to form cocrystals with coformers that exhibit the functional groups observed in the CSD survey, including our drug coformer candidates.

Table 2. Ranking positions for FUR coformers reported at CSD, including the two drugs used in this study (in bold) based on COSMOQuick calculations. Non-drug molecules marked with *.

Coformer	H_{ex}(kcal/mol)	Ref. for the Corresponding Cocrystal/Salt
1,10-phenanthroline *	−5.462215	[33]
4,4′-bipyridine *	−3.87421	[34]
Piperazine *	−3.85138	[9]
Triamterene	−3.33838	[29]
Pentoxifylline	−3.18718	[8]
Cytosine *	−3.018425	[30]
Caffeine	−2.910815	[30]
Gefitinib	−2.8597	[11]
4-Aminopyridine *	−2.6074	[35]
Urea *	−2.41366	[36]
Ethenzamide	**−2.36084**	**This work**
Erlotinib	−2.3474	[10]
Nicotinamide	−2.13354	[37]
5-fluorocytosine	−2.101	[12]
4-toluamide *	−1.95134	[9]
2,2′-bipyridine *	−1.83637	[33]
2-picolinamide *	−1.72857	[9]
Anthranilamide *	−1.36987	[9]
Piroxicam	**−0.91377**	**This work**, [31] **for acetone solvate**

3.2. Mechanochemical Synthesis

Mechanochemistry has proved to be a powerful tool to obtain multi-component solid forms (salts, cocrystals, hydrates/solvates and their respective combinations), particularly in searching for new solids involving pharmaceuticals [38–41]. Cocrystallization of FUR with the corresponding coformers was carried out using various stoichiometries (1:1, 1:2, and 2:1). The patterns obtained by grinding different molar ratios of the two components were compared with the patterns of isolated API and coformers. The comparison shows that all three ratios have common characteristic peaks that were different from the two APIs. The 1:2 and 2:1 FUR:coformer patterns also contained peaks characteristics to one of the components (Figure S1, in Supplementary Materials). Only the 1:1 products had a completely different pattern where all reflections of the reagents disappeared completely, thus revealing new phases. These polycrystalline materials were used for further recrystallization to obtain suitable crystals for structure determination. In addition, there is a good agreement between the experimental and the simulated patterns (Figures S2 and S3). This synthetic approach prevented the risk for solvate formation in the case of FUR-PRX. An acetone solvate was reported previously [31] as having been obtained from acetone solution of a 1:1 stoichiometric mixture of FUR and PRX by slow evaporation.

3.3. Structural Studies of Multi-Component Forms

Single-crystal X-ray diffraction analysis (Table 1 and Figures S4 and S5) confirmed the cocrystal nature of FUR–ETZ and FUR–PRX obtained by LAG of the APIs in methanol.

Figure 1 shows a PXRD overlay of the ground and starting materials and the simulated pattern from the single crystal structure. This figure shows that the ground material matches the one from single crystal analysis, corresponding to FUR–ETZ cocrystal.

FUR–ETZ cocrystal crystallized in the monoclinic $P2_1/c$ space group. The asymmetric unit was composed of FUR and ETZ in a 1:1 stoichiometric ratio (Figure 2a). The cocrystal adopts a ribbon structure through the acid···amide synthons which connect FUR catemer-like chains formed from $SO_2HN-H\cdots O_{furan}$ weak hydrogen bonds between neighboring FUR molecules (Figure 2b). The ribbons stack 6-membered aromatic rings of FUR and ETZ (centroid–centroid distance: 3.7262(17) Å) to form columns running along the b axis (Figure 2c). These columns are reinforced by H-bonding interactions involving the sulfonamide and amide moieties of the cocrystal. Finally, weak C–H···O$_{sulfonamide}$ hydrogen bonds connect these columns to form the 3D structure.

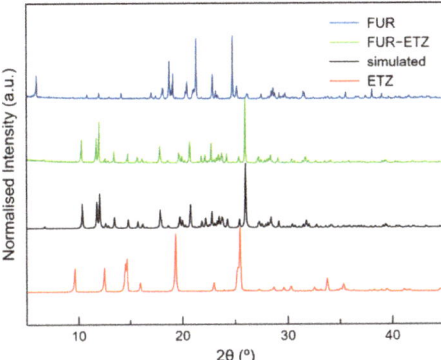

Figure 1. PXRD patterns of the new phase FUR–ETZ obtained by liquid-assisted grinding (LAG) with methanol (MET) solvent, the simulated pattern from crystal structure and the corresponding reactants.

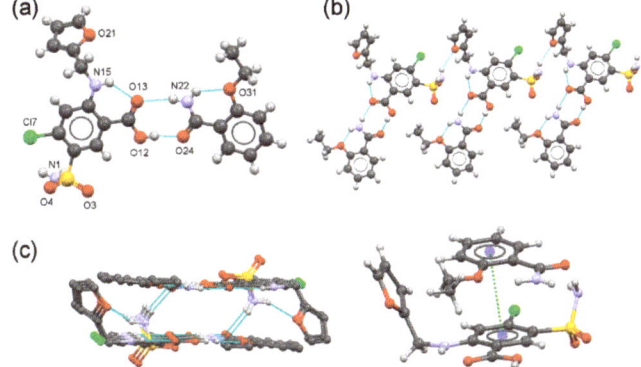

Figure 2. (a) Asymmetric unit of the FUR–ETZ cocrystal. (b) acid···amide and sulfonamide···furan synthons give a ribbon along the b-axis by H-bonding interactions (Table S1). (c) Left. Detail of the column structure in FUR–ETZ. Carbon bound H atoms omitted for clarity Right. π–π stacking interaction in the FUR–ETZ cocrystal.

Figure 3 shows a PXRD overlay of the ground and starting materials, as well as the simulated patterns from the single crystal structure and the reported acetone solvate [31]. As shown in this figure, the ground material matches a new single crystal phase, corresponding to the FUR–PRX solid form.

FUR–PRX cocrystal crystallizes in the monoclinic space group $P2_1/n$. The crystal structure contains one molecule each of FUR and PRX in the asymmetric unit that are associated by the heterosynthon acid···pyridine (Figure 4a). PRX molecules exhibits a strong intramolecular H-bonding interaction O-H···O=C (Table S3). FUR molecules form centrosymmetric dimers through H-bonding interactions involving sulfonamide groups (SO$_2$HN–H···O=S) and connect PRX molecules by additional H-bonds (SO$_2$HN–H···O=S$_{PRX}$) to generate ribbons running along a axis. The ribbons have FUR dimers forming the backbone of the ribbon and PRX molecules in the periphery (Figure 4b). The structure is additionally stabilized by weak C–H···O hydrogen bonds formed from sulfonamide oxygen atoms with methyl groups from PRX molecules to form the 3D structure.

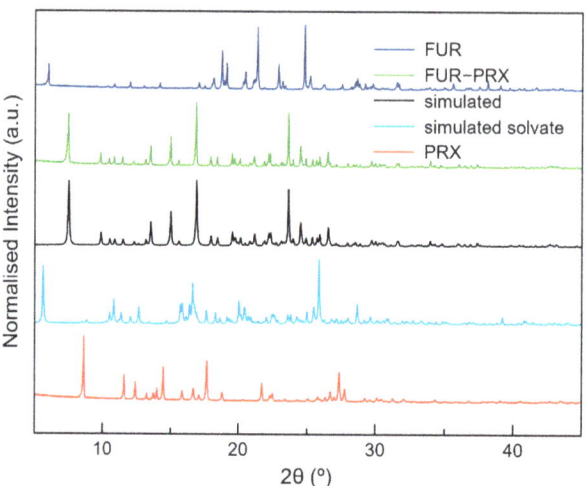

Figure 3. PXRD patterns of the new phase FUR–PRX obtained by liquid-assisted grinding (LAG) with methanol (MET) solvent, the simulated patterns from crystal structure and reported acetone solvate structure and the corresponding reactants.

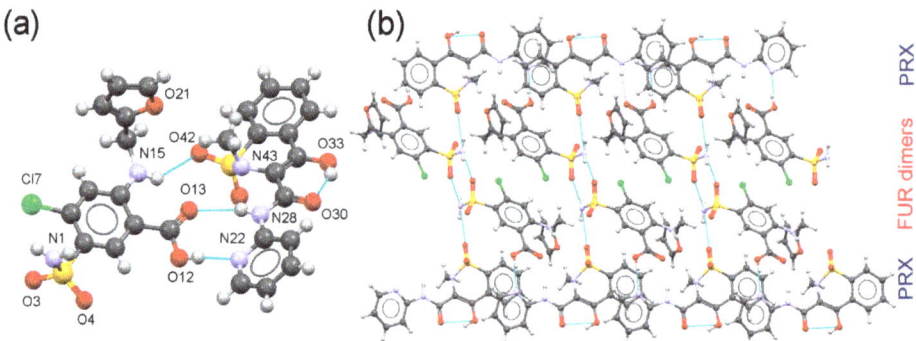

Figure 4. (**a**) Asymmetric unit of the FUR–PRX cocrystal. (**b**) Detail of the ribbon structure along the *a* axis.

All the reported polymorphs of FUR exhibit carboxylic dimer synthons; however, each polymorph has a variation in the hydrogen bonding of sulfonamide groups giving different synthons. In the stable FUR polymorph 1 [42], a robust dimeric centrosymmetric H-bonding interaction between sulfonamide groups is observed that further generate a linear tape structure. As expected, in both drug–drug FUR cocrystals, carboxylic dimer synthon is disrupted by the insertion of the amide or pyridine functional group for ETZ or PRX coformer, respectively. Moreover, in the case of FUR-ETZ, the sulfonamide synthon observed in the FUR polymorph 1 is replaced by two different synthons involving FUR and ETZ meanwhile in FUR-PRX, this synthon is partially maintained as sulfonamide dimer but the linear tape structure is blocked by PRX molecules. The resulting ribbon structures are different in both cocrystals and in principle would anticipate that both cocrystal will exhibit different physicochemical properties as will be discussed in the following sections.

3.4. Fourier Transform Infrared (FT-IR) Spectroscopy

FT–IR is a helpful technique that quickly detects the formation of novel multi-component pharmaceutical solid forms [43]. Changes in vibrational frequencies due to cocrystal/salt formation can be easily monitored. When the two APIs are joined together in the solid form, the reported IR bands with diagnostic values are expected to be shifted, thus indicating the presence of intermolecular forces between functional groups—i.e., hydrogen bonds—which build the cocrystal structures [44]. Band assignments (Table 3) were performed based on the crystallographic analysis (Section 3.1) and considering the spectroscopic data available for related FUR compounds found in the literature [13].

Table 3. Summary of relevant FT–IR vibrational frequencies (cm^{-1}) in the spectra of FUR, FUR–ETZ, and FUR–PRX.

Compound	ν(NH$_2$) Sulfonamide	ν(NH) Secondary Amine	ν(C=O) Carboxyl	ν(COO$^-$) Carboxylate	ν(S=O) Sulfonamide
FUR	(as) 3400 (s) 3351	3285	1670	-	(as) 1328 (s) 1139
FUR–ETZ	(as) 3438 (s) 3291	3285	1670	-	(as) 1339 (s) 1154
FUR–PRX	(as) 3317 (s) 3230	3269	1670	-	(as) 1339 (s) 1154

FUR exhibits stretching frequencies at 3400 and 3351 cm^{-1} (sulfonamide primary amine), 3285 cm^{-1} (sulfonamide secondary amine), 1670 cm^{-1} (carboxyl stretch), and 1328 and 1139 cm^{-1} (sulfonamide S=O stretching modes). The FT-IR spectra of FUR and the multi-component forms are shown in Figure 5. In FUR–ETZ and FUR–PRX cocrystals the band corresponding to carboxyl group (1670 cm^{-1}) appears in the same position as in FUR. In the cocrystals, the –NH$_2$ asymmetric and symmetric stretching modes are shifted (3438 and 3291 cm^{-1} for FUR–ETZ and 3317 and 3230 cm^{-1} for FUR–PRX). S=O stretching modes are shifted to 1345 and 1143 cm^{-1} in FUR–ETZ, 1339 and 1154 cm^{-1} in the case of FUR–PRX, confirming that these functional groups interact with the coformer, as demonstrated in the crystal structures analysis. The FT–IR vibrational frequency comparisons are summarized in Table 3.

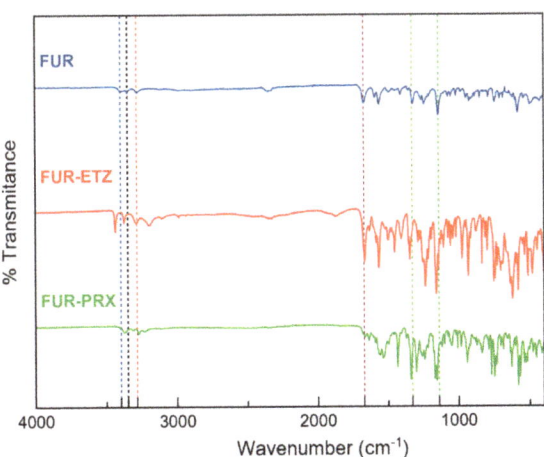

Figure 5. Comparison of Fourier transform infrared (FT−IR) spectra of FUR, FUR–ETZ and FUR–PRX solid forms.

3.5. Thermal Analysis

It is well accepted that the melting point of an API can be altered through cocrystallization [45]. The outcome will generally be a solid with a melting point between (M), lower (L), or higher (H) than the isolated API and coformer, following the occurrence trend M >> L > H [46]. The thermal behavior of the reported compounds was studied by DSC. In Figure 6, the DSC of the corresponding FUR–ETZ and FUR–PRX cocrystals are reported. Each trace shows one single endothermic event, which represents the melting point of these pure species. Interestingly, while the melting point of the FUR–ETZ cocrystal (187.67 °C) is in between those of the reported for the two reference APIs (ETZ: 129–134 °C; FUR: 203–205 °C), FUR–PRX cocrystal shows a melting endotherm at 214.82 °C, higher than the melting point of its components (PRX: 201.89 °C; FUR: 203–205 °C), an indication that this pharmaceutical cocrystal is thermally more stable than FUR by itself. Although the density and packing coefficient of the cocrystals are similar (Table 1), the overall packing arrangement of FUR-PRX and the non-covalent interactions involved impact its thermal behavior.

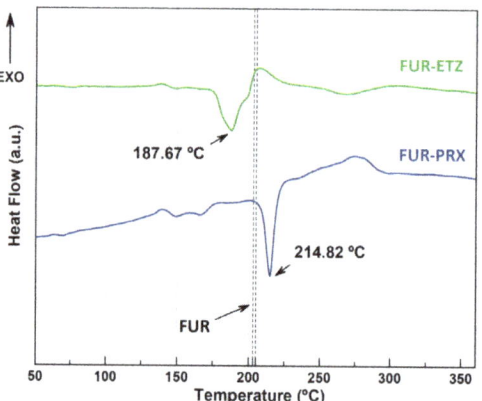

Figure 6. Differential scanning calorimetry (DSC) plots of FUR–ETZ and FUR–PRX. Dotted lines correspond to the range of temperature reported for melting of FUR.

3.6. Stability Studies

The stability of cocrystals was studied in this work by performing aqueous slurry experiments at 25 °C and storing them at accelerated ageing conditions (40 °C and 75% relative humidity). The thermodynamic stability of cocrystals was first evaluated by slurry experiments at 25 °C. In these experiments, excess solids of the cocrystal powders were stirred in deionized water for 24 h. The resulting filtered and air-dried samples were analyzed by PXRD to evaluate their phase purity, and it was observed that the two cocrystals were stable upon slurrying. These observations suggest that the cocrystals are thermodynamically stable at room temperature. Likewise, results of the stability tests suggest that the two new solid forms remained the same after storage for two months (Figure 7). The stability of the cocrystals at accelerated test conditions is consistent with the thermodynamic stability observed in the slurry experiments.

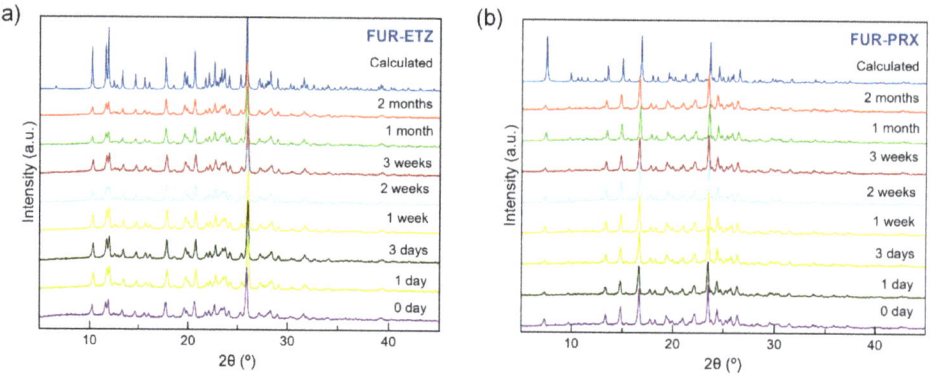

Figure 7. PXRD patterns of FUR–ETZ (**a**) and FUR–PRX (**b**) with respect to the stability under accelerated ageing conditions (40 °C, 75% RH) at different time intervals.

3.7. Equilibrium Solubility

As observed in the previous section, cocrystals were thermodynamically stable when suspended in water at room temperature. They did not transform to component phases or there is no evidence of phase transitions, suggesting that these phases have equal or lower solubility than the drug or coformer. The equilibrium solubility of the new FUR cocrystals was lower than that of FUR at pH 7.4. As seen in Table 4, cocrystals possess lower solubility than FUR. Although the differences in equilibrium solubility between FUR and the cocrystals were significant, interestingly, the extent of enhanced solubility of the multi-component solids is more significant than the solubility of the drug coformers.

Table 4. Equilibrium solubility of FUR and its cocrystals in water PBS pH 7.4.

Solid Form	Equilibrium Solubility at 25 °C (mg/mL)	Extent of Increase Relative to the Solubility of FUR.	Extent of Increase Relative to the Solubility of Coformer.
FUR	2.40	-	-
FUR–ETZ	1.24	×0.52	×41 (ETZ) [a]
FUR–PRX	1.47	×0.61	×84 (PRX) [b]

[a] Reported solubility of ETZ at 27 °C: 0.03 mg/mL [47]. [b] Reported solubility of PRX at 37 °C: 0.0198 mg/mL [48].

4. Conclusions

In conclusion, we have described two new drug–drug cocrystals containing FUR and ETZ and PRX as coformers. A mechanochemical synthetic route have allowed to avoid hydrate/solvate formation as evidenced in the case of FUR-PRX. Expected hydrogen bonds contributed by the drug coformers sustain the cocrystals, disrupting the acid:aciddimer synthon observed in the stable FUR polymorph 1. All the solids exhibit good thermal stability, and good stability under accelerated ageing. Although they do not exhibit increased solubility than FUR drug, the drug coformers notably do. However, the success of these drug–drug cocrystals as potential fixed-dose solids requires an appropriately designed clinical study to establish their safety and effectiveness.

Supplementary Materials: The following are available online at https://www.mdpi.com/article/10.3390/cryst11111339/s1, Figure S1. PXRD patterns of FUR–ETZ obtained by grinding the two components at different molar ratios. Blue dotted lines indicate characteristic FUR reflections. Orange dotted lines indicate characteristic ETZ reflections; Figure S2. PXRD patterns of FUR–PRX obtained by grinding the two components at different molar ratios. Blue dotted lines indicate characteristic FUR reflections. Orange dotted lines indicate characteristic PRX reflections; Figure S3. Le bail profile fit (red line) to the experimental PXRD data (blue line) of FUR-ETZ (a) and FUR-PRX (b). The profile fitting for both the cocrystals shows low discrepancy (grey line); Figure S4. ORTEP representation showing the asymmetric unit of FUR—ETZ with atom numbering scheme (thermal ellipsoids are plotted with the 50% probability level); Figure S5. ORTEP representation showing the asymmetric unit of FUR—PRX with atom numbering scheme (thermal ellipsoids are plotted with the 50% probability level); Figure S6. TGA traces of FUR–ETZ (top) and FUR–PRX (bottom); Figure S7. Solubility curve of FUR—ETZ in water PBS at pH 7.4; Figure S8. Solubility curve of FUR—PRX in water PBS at pH 7.4; Figure S9. PXRD patterns of FUR–ETZ after the stability slurry assay (at 25 °C, during 24 h, in water); Figure S10. PXRD patterns of FUR–PRX after the stability slurry assay (at 25 °C, during 24 h, in water); Table S1. Hydrogen bonds for FUR—ETZ (Å and deg.); Table S2. π,π-stacking interactions analysis of compound FUR—ETZ. Table S3. Hydrogen bonds for FUR—PRX (Å and deg.).

Author Contributions: Conceptualization and methodology, D.C.-L.; Formal analysis and investigation, C.A.-P., L.R.-D., F.J.A.-M., A.D.-M. and J.G.-M.; Writing—original draft preparation, D.C.-L.; Writing—review and editing, D.C.-L.; Funding acquisition, D.C.-L. and J.G.-M.; Supervision, D.C.-L. All authors have read and agreed to the published version of the manuscript.

Funding: This research was funded by Spanish Agencia Estatal de Investigación of the Ministerio de Ciencia, Innovación y Universidades (MICIU) and co-funded with FEDER, UE, Project No. PGC2018-102047-B-I00 (MCIU/AEI/FEDER, UE) and Project No. B-FQM-478-UGR20 (FEDER-Universidad de Granada, Spain).

Institutional Review Board Statement: Not applicable.

Informed Consent Statement: Not applicable.

Data Availability Statement: Not applicable.

Acknowledgments: F.J.A.-M. wants to acknowledge an FPI grant (ref. PRE2019-088832).

Conflicts of Interest: The authors declare no conflict of interest.

References

1. Jackson, E.K. Diuretics. In *Goodman and Gilman's The Pharmacological Basis of Therapeutics*; Brunton, L., Lazo, J., Parker, K., Eds.; McGraw-Hill: New York, NY, USA, 2006; pp. 737–770. ISBN 0071422803.
2. Carone, L.; Oxberry, S.G.; Twycross, R.; Charlesworth, S.; Mihalyo, M.; Wilcock, A. Furosemide. *J. Pain Symptom Manag.* **2016**, *52*, 144–150. [CrossRef]
3. Amidon, G.L.; Lennernäs, H.; Shah, V.P.; Crison, J.R. A Theoretical Basis for a Biopharmaceutic Drug Classification: The Correlation of in Vitro Drug Product Dissolution and in Vivo Bioavailability. *Pharm. Res.* **1995**, *12*, 413–420. [CrossRef]
4. Wisher, D. Martindale: The Complete Drug Reference. 37th Ed. *J. Med Libr. Assoc. JMLA* **2012**, *100*, 75–76. [CrossRef]
5. Grahnén, A.; Hammarlund, M.; Lundqvist, T. Implications of Intraindividual Variability in Bioavailability Studies of Furosemide. *Eur. J. Clin. Pharmacol.* **1984**, *27*, 595–602. [CrossRef]
6. Berry, D.J.; Steed, J.W. Pharmaceutical Cocrystals, Salts and Multicomponent Systems; Intermolecular Interactions and Property Based Design. *Adv. Drug Deliv. Rev.* **2017**, *117*, 3–24. [CrossRef] [PubMed]
7. Kumar, A.; Kumar, S.; Nanda, A. A Review about Regulatory Status and Recent Patents of Pharmaceutical Co-Crystals. *Adv. Pharm. Bull.* **2018**, *8*, 355–363. [CrossRef]
8. Stepanovs, D.; Mishnev, A. Multicomponent Pharmaceutical Cocrystals: Furosemide and Pentoxifylline. *Acta Crystallogr. Sect. C Cryst. Struct. Commun.* **2012**, *68*, o488–o491. [CrossRef] [PubMed]
9. Banik, M.; Gopi, S.P.; Ganguly, S.; Desiraju, G.R. Cocrystal and Salt Forms of Furosemide: Solubility and Diffusion Variations. *Cryst. Growth Des.* **2016**, *16*, 5418–5428. [CrossRef]
10. George, C.P.; Thorat, S.H.; Shaligram, P.S.; Suresha, P.R.; Gonnade, R.G. Drug-Drug Cocrystals of Anticancer Drugs Erlotinib-Furosemide and Gefitinib-Mefenamic Acid for Alternative Multi-Drug Treatment. *CrystEngComm* **2020**, *22*, 6137–6151. [CrossRef]
11. Thorat, S.H.; Sahu, S.K.; Patwadkar, M.V.; Badiger, M.V.; Gonnade, R.G. Drug-Drug Molecular Salt Hydrate of an Anticancer Drug Gefitinib and a Loop Diuretic Drug Furosemide: An Alternative for Multidrug Treatment. *J. Pharm. Sci.* **2015**, *104*, 4207–4216. [CrossRef] [PubMed]
12. Diniz, L.F.; Carvalho, P.S.; Pena, S.A.C.; Gonçalves, J.E.; Souza, M.A.C.; de Souza Filho, J.D.; Bomfim Filho, L.F.O.; Franco, C.H.J.; Diniz, R.; Fernandes, C. Enhancing the Solubility and Permeability of the Diuretic Drug Furosemide via Multicomponent Crystal Forms. *Int. J. Pharm.* **2020**, *587*, 119694. [CrossRef] [PubMed]
13. Abraham Miranda, J.; Garnero, C.; Chattah, A.K.; Santiago De Oliveira, Y.; Ayala, A.P.; Longhi, M.R. Furosemide:Triethanolamine Salt as a Strategy to Improve the Biopharmaceutical Properties and Photostability of the Drug. *Cryst. Growth Des.* **2019**, *19*, 2060–2068. [CrossRef]
14. Wang, X.; Du, S.; Zhang, R.; Jia, X.; Yang, T.; Zhang, X. Drug-Drug Cocrystals: Opportunities and Challenges. *Asian J. Pharm. Sci.* **2021**, *16*, 307–317. [CrossRef]
15. Herchuelz, A.; Derenne, F.; Deger, F.; Juvent, M.; van Ganse, E.; Staroukine, M.; Verniory, A.; Boeynaems, J.M.; Douchamps, J. Interaction between Nonsteroidal Anti-Inflammatory Drugs and Loop Diuretics: Modulation by Sodium Balance. *J. Pharmacol. Exp. Ther.* **1989**, *248*, 1175–1181.
16. Paterson, C.A.; Jacobs, D.; Rasmussen, S.; Youngberg, S.P.; McGuinness, N. Randomized, Open-Label, 5-Way Crossover Study to Evaluate the Pharmacokinetic/Pharmacodynamic Interaction between Furosemide and the Non-Steroidal Anti-Inflammatory Drugs Diclofenac and Ibuprofen in Healthy Volunteers. *Int. J. Clin. Pharmacol. Ther.* **2011**, *49*, 477–490. [CrossRef]
17. Moore, N.; Pollack, C.; Butkerait, P. Adverse Drug Reactions and Drug–Drug Interactions with over-the-Counter NSAIDs. *Ther Clin Risk Manag.* **2015**, *11*, 1061–1075. [CrossRef]
18. Baker, D.E. Piroxicam—Furosemide Drug Interaction. *Drug Intell. Clin. Pharm.* **1988**, *22*, 505–506. [CrossRef] [PubMed]
19. Allen, F.H. The Cambridge Structural Database: A Quarter of a Million Crystal Structures and Rising. *Acta Crystallogr. Sect. B Struct. Sci.* **2002**, *58*, 380–388. [CrossRef] [PubMed]
20. Loschen, C.; Klamt, A. Solubility Prediction, Solvate and Cocrystal Screening as Tools for Rational Crystal Engineering. *J. Pharm. Pharmacol.* **2015**, *67*, 803–811. [CrossRef]
21. Coelho, A.A. TOPAS and TOPAS-Academic: An Optimization Program Integrating Computer Algebra and Crystallographic Objects Written in C++: An. *J. Appl. Crystallogr.* **2018**, *51*, 210–218. [CrossRef]
22. Bruker APEX3. *APEX3 V2019.1*; Bruker-AXS: Madison, WI, USA, 2019.
23. Sheldrick, G.M. SHELXT—Integrated Space-Group and Crystal-Structure Determination. *Acta Crystallogr. Sect. A Found. Crystallogr.* **2015**, *71*, 3–8. [CrossRef] [PubMed]
24. Sheldrick, G.M. Crystal Structure Refinement with SHELXL. *Acta Crystallogr. Sect. C Struct. Chem.* **2015**, *71*, 3–8. [CrossRef]

25. Dolomanov, O.V.; Bourhis, L.J.; Gildea, R.J.; Howard, J.A.K.; Puschmann, H. OLEX2: A Complete Structure Solution, Refinement and Analysis Program. *J. Appl. Crystallogr.* **2009**, *42*, 339–341. [CrossRef]
26. Spek, A.L. Structure Validation in Chemical Crystallography. *Acta Crystallogr. Sect. D Biol. Crystallogr.* **2009**, *65*, 148–155. [CrossRef]
27. Macrae, C.F.; Bruno, I.J.; Chisholm, J.A.; Edgington, P.R.; McCabe, P.; Pidcock, E.; Rodriguez-Monge, L.; Taylor, R.; van de Streek, J.; Wood, P.A. Mercury CSD 2.0—New Features for the Visualization and Investigation of Crystal Structures. *J. Appl. Crystallogr.* **2008**, *41*, 466–470. [CrossRef]
28. Horst, J.H.T.; Deij, M.A.; Cains, P.W. Discovering New Co-Crystals. *Cryst. Growth Des.* **2009**, *9*. [CrossRef]
29. Peng, B.; Wang, J.R.; Mei, X. Triamterene–Furosemide Salt: Structural Aspects and Physicochemical Evaluation. *Acta Crystallogr. Sect. B: Struct. Sci. Cryst. Eng. Mater.* **2018**, *74*, 738–741. [CrossRef]
30. Goud, N.R.; Gangavaram, S.; Suresh, K.; Pal, S.; Manjunatha, S.G.; Nambiar, S.; Nangia, A. Novel Furosemide Cocrystals and Selection of High Solubility Drug Forms. *J. Pharm. Sci.* **2012**, *101*, 664–680. [CrossRef]
31. Mishnev, A.; Kiselovs, G. New Crystalline Forms of Piroxicam. *Z. Fur Nat.-Sect. C J. Biosci.* **2013**, *68 B*, 168–174. [CrossRef]
32. Abramov, Y.A.; Loschen, C.; Klamt, A. Rational Coformer or Solvent Selection for Pharmaceutical Cocrystallization or Desolvation. *J. Pharm. Sci.* **2012**, *101*, 3687–3697. [CrossRef] [PubMed]
33. Sangtani, E.; Mandal, S.K.; Sreelakshmi, A.S.; Munshi, P.; Gonnade, R.G. Salts and Cocrystals of Furosemide with Pyridines: Differences in π-Stacking and Color Polymorphism. *Cryst. Growth Des.* **2017**, *17*, 3071–3087. [CrossRef]
34. Srirambhatla, V.K.; Kraft, A.; Watt, S.; Powell, A.V. A Robust Two-Dimensional Hydrogen-Bonded Network for the Predictable Assembly of Ternary Co-Crystals of Furosemide. *CrystEngComm* **2014**, *16*, 9979–9982. [CrossRef]
35. Sangtani, E.; Sahu, S.K.; Thorat, S.H.; Gawade, R.L.; Jha, K.K.; Munshi, P.; Gonnade, R.G. Furosemide Cocrystals with Pyridines: An Interesting Case of Color Cocrystal Polymorphism. *Cryst. Growth Des.* **2015**, *15*, 5858–5872. [CrossRef]
36. Rahal, O.; Majumder, M.; Spillman, M.J.; van de Streek, J.; Shankland, K. Co-Crystal Structures of Furosemide:Urea and Carbamazepine:Indomethacin Determined from Powder x-Ray Diffraction Data. *Crystals* **2020**, *10*, 42. [CrossRef]
37. Ueto, T.; Takata, N.; Muroyama, N.; Nedu, A.; Sasaki, A.; Tanida, S.; Terada, K. Polymorphs and a Hydrate of Furosemide-Nicotinamide 1:1 Cocrystal. *Cryst. Growth Des.* **2012**, *12*, 485–494. [CrossRef]
38. Braga, D.; Maini, L.; Grepioni, F. Mechanochemical Preparation of Co-Crystals. *Chem. Soc. Rev.* **2013**, *42*, 7638–7648. [CrossRef]
39. Delori, A.; Friščić, T.; Jones, W. The Role of Mechanochemistry and Supramolecular Design in the Development of Pharmaceutical Materials. *CrystEngComm* **2012**, *14*, 2350. [CrossRef]
40. Friščić, T.; Childs, S.L.; Rizvi, S.A.A.; Jones, W. The Role of Solvent in Mechanochemical and Sonochemical Cocrystal Formation: A Solubility-Based Approach for Predicting Cocrystallisation Outcome. *CrystEngComm* **2009**, *11*, 418–426. [CrossRef]
41. Verdugo-Escamilla, C.; Alarcón-Payer, C.; Frontera, A.; Acebedo-Martínez, F.J.; Domínguez-Martín, A.; Gómez-Morales, J.; Choquesillo-Lazarte, D. Interconvertible Hydrochlorothiazide–Caffeine Multicomponent Pharmaceutical Materials: A Solvent Issue. *Crystals* **2020**, *10*, 1088. [CrossRef]
42. Babu, N.J.; Cherukuvada, S.; Thakuria, R.; Nangia, A. Conformational and Synthon Polymorphism in Furosemide (Lasix). *Cryst. Growth Des.* **2010**, *10*, 1979–1989. [CrossRef]
43. Heinz, A.; Strachan, C.J.; Gordon, K.C.; Rades, T. Analysis of Solid-State Transformations of Pharmaceutical Compounds Using Vibrational Spectroscopy. *J. Pharm. Pharmacol.* **2009**, *61*, 971–988. [CrossRef] [PubMed]
44. Mukherjee, A.; Tothadi, S.; Chakraborty, S.; Ganguly, S.; Desiraju, G.R. Synthon Identification in Co-Crystals and Polymorphs with IR Spectroscopy. Primary Amides as a Case Study. *CrystEngComm* **2013**, *15*, 4640–4654. [CrossRef]
45. Schultheiss, N.; Newman, A. Pharmaceutical Cocrystals and Their Physicochemical Properties. *Cryst. Growth Des.* **2009**, *9*, 2950–2967. [CrossRef] [PubMed]
46. Perlovich, G. Melting Points of One- and Two-Component Molecular Crystals as Effective Characteristics for Rational Design of Pharmaceutical Systems. *Acta Crystallogr. Sect. B Struct. Sci. Cryst. Eng. Mater.* **2020**, *76*, 696–706. [CrossRef]
47. Khatioda, R.; Bora, P.; Sarma, B. Trimorphic Ethenzamide Cocrystal: In Vitro Solubility and Membrane Efflux Studies. *Cryst. Growth Des.* **2018**, *18*, 4637–4645. [CrossRef]
48. Karataş, A.; Yüksel, N.; Baykara, T. Improved Solubility and Dissolution Rate of Piroxicam Using Gelucire 44/14 and Labrasol. *Farmaco* **2005**, *60*, 777–782. [CrossRef]

Article

Multicomponent Materials to Improve Solubility: Eutectics of Drug Aminoglutethimide

Basanta Saikia *, Andreas Seidel-Morgenstern and Heike Lorenz

Max Planck Institute for Dynamics of Complex Technical Systems, 39106 Magdeburg, Germany; seidel@mpi-magdeburg.mpg.de (A.S.-M.); lorenz@mpi-magdeburg.mpg.de (H.L.)
* Correspondence: saikia@mpi-magdeburg.mpg.de

Abstract: Here, we report the synthesis and experimental characterization of three drug-drug eutectic mixtures of drug aminoglutethimide (AMG) with caffeine (CAF), nicotinamide (NIC) and ethenzamide (ZMD). The eutectic mixtures i.e., AMG-CAF (1:0.4, molar ratio), AMG-NIC (1:1.9, molar ratio) and AMG-ZMD (1:1.4, molar ratio) demonstrate significant melting point depressions ranging from 99.2 to 127.2 °C compared to the melting point of the drug AMG (151 °C) and also show moderately higher aqueous solubilities than that of the AMG. The results presented include the determination of the binary melt phase diagrams and accompanying analytical characterization via X-ray powder diffraction, FT-IR spectroscopy and scanning electron microscopy.

Keywords: aminoglutethimide; eutectic mixture; solubility; melt phase diagram

Citation: Saikia, B.; Seidel-Morgenstern, A.; Lorenz, H. Multicomponent Materials to Improve Solubility: Eutectics of Drug Aminoglutethimide. *Crystals* **2022**, *12*, 40. https://doi.org/10.3390/cryst12010040

Academic Editors: Duane Choquesillo-Lazarte and Alicia Dominguez-Martin

Received: 29 November 2021
Accepted: 24 December 2021
Published: 28 December 2021

Publisher's Note: MDPI stays neutral with regard to jurisdictional claims in published maps and institutional affiliations.

Copyright: © 2021 by the authors. Licensee MDPI, Basel, Switzerland. This article is an open access article distributed under the terms and conditions of the Creative Commons Attribution (CC BY) license (https://creativecommons.org/licenses/by/4.0/).

1. Introduction

The multicomponent solid form of the drug is rapidly emerging as an effective way to improve the drug physiochemical properties such as solubility, dissolution rate, bioavailability and other crucial pharmaceutical properties like stability, hygroscopicity, chemical stability, flowability, etc. [1–7]. Eutectic mixtures are multicomponent compounds made up of two or more crystalline solids that show immiscibility in the solid-state and do not combine to generate a new chemical compound but, at a certain ratio, the eutectic composition will exhibit a melting or solidification point significantly lower than its constituents [1,4,8]. The formation of eutectic mixtures can occur via different noncovalent interactions primarily hydrogen bonding, van der Waals forces, and aromatic interactions etc. [9,10]. It is a trial and error approach to get a eutectic composition for enhancing the solubility and bioavailability of medicines with limited water solubility in BCS classes II and IV [1,4]. Eutectic mixtures are frequently used for the design of medicines and delivery methods for administration routes [1,11]. For example, the eutectic mixture of lidocaine and prilocaine cream is a novel formulation of dermal anesthesia which is effective and safe for the treatment option in premature ejaculation patients of various types [11]. Eutectic mixtures generally exhibit high thermodynamic parameters, for example, free energy, enthalpy and entropy etc., which alter the solubility and dissolution behavior [1]. Besides, a higher soluble eutectic mixture component, called coformer in the following, also can favorably influence the wettability of the drug, thus improving the bioavailability [12,13]. When taken orally, curcumin, for example, has low bioavailability and solubility. However, the eutectic mixture of curcumin and nicotinamide in a 1:2 ratio exhibits a 10-fold faster intrinsic dissolution rate [14]. Another aspect is that it is critical to identifying the production of eutectics during the formulation stage to minimize manufacturing difficulties [15,16]. During pharmaceutical research, understanding eutectic mixtures can aid in the discovery of compounds with equivalent melting points.

In this work, we studied the chiral drug aminoglutethimide [3-(4-aminophenyl)-3-ethyl-2, 6-piperidinedione] (AMG), a nonsteroidal aromatase inhibitor drug, used for the treatment of Cushing's syndrome, breast cancer, and prostate cancer [17–19]. According to the Biopharmaceutics Classification System (BCS), AMG is classified as a BCS class II drug because of its poor solubility [20]. Attempts were made in this work to create multidrug eutectics of AMG considering caffeine (CAF, a psychoactive drug [21]), nicotinamide (NIC, a vitamin [22]) and ethenzamide (ZMD, nonsteroidal anti-inflammatory drug [23]) as coformers to improve the aqueous solubility of AMG. Scheme 1 shows the chemical structures of AMG and the three coformers. For the synthesis of eutectics, mainly a mechanochemical solvent-assisted grinding (LAG) method has been used. Differential scanning calorimetry (DSC) data is utilized to establish the precise eutectic composition. X-ray powder diffraction (PXRD), Fourier-transform infrared spectroscopy (FT-IR) and scanning electron microscopy (SEM) were applied to characterize the eutectic mixtures and their components. Further, the aqueous solubility of the eutectics was determined and found to moderately exceed that of the parent compound AMG.

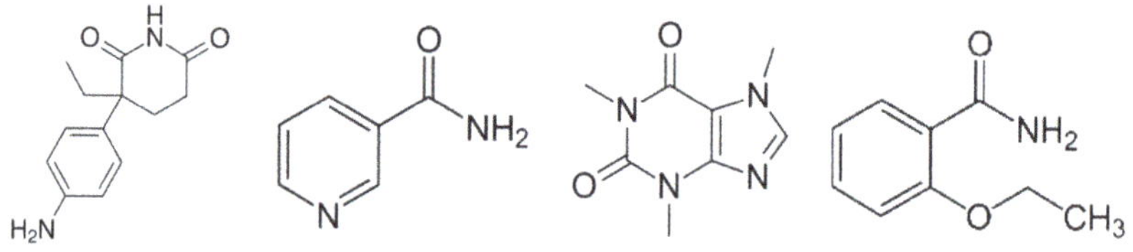

Scheme 1. Molecular structure of Aminoglutethimide and the coformers selected for the study.

2. Materials and Methods

2.1. Materials

Aminoglutethimide (purity: >98.0%) was purchased from TCI, Japan. Caffeine (purity: >98%), nicotinamide (purity: >98%) and ethenzamide (purity: >97%) were purchased from Sigma Aldrich (Darmstadt, Germany). Millipore water from the Milli-Q system (Merck Millipore, Milli-Q Advantage, Darmstadt, Germany) was used for solubility determination and HPLC-grade solvents for the mechanochemical experiments.

2.2. Aminoglutethimide Eutectic Mixture Screening and Eutectic Composition Determination

Multiple mixtures of AMG and the chosen coformer in different weight percentages, e.g., 50%, 55%, 60%, 65%, 70%, 75%, and 80%, were prepared. The needed amount of each component was added to a mortar to obtain 100 mg of the desired binary combination and grinded using mortar and pestle with a dropwise addition of acetonitrile for 30 min.

The melting behavior of the resulting solids and the eutectic formation was determined by means of a DSC linear heating run. Eutectic and pure compounds melting temperatures were considered from the corresponding peak onsets, and liquidus temperatures for mixtures were taken from the peak maximum. Therewith, the binary phase diagrams and corresponding Tammann plots were constructed to specify the eutectic composition of the respective drug-drug system. For the Tammann plot, the enthalpy of fusion of the eutectic melting effects in the DSC curves of the mixtures are used.

The melting of the eutectic mixture is characterized by a single melt peak in the DSC curve for the corresponding drug-drug, that is an eutectic composition. This is because the eutectic is an invariant point in the binary system in analogy to the melting point of a pure compound in a unary system.

2.3. Preparation of Bulk Mixtures at the Eutectic Composition

For liquid-assisted grinding, the components of each system were mixed and homogenized for 30 min in a glass mortar and pestle with acetonitrile as the solvent. The AMG-CAF, AMG-NIC and AMG-ZMD systems were produced in their respective eutectic compositions to perform solid-state characterization and solubility tests. In a nutshell, exact weights of AMG-CAF (75 wt.% of AMG), AMG-NIC (50 wt.% of AMG), and AMG-ZMD (50 wt.% of AMG) systems were considered for the solubility evaluation.

2.4. Analytical Techniques

Powder X-ray Diffraction (PXRD). The solid-state properties of the system were investigated by powder X-ray diffraction (PXRD) analysis with an X'pert Pro Diffractometer (PANalytical GmbH, Kassel, Germany) with Cu Kα radiation and an X'Celerator detector in the 2-theta range of 3–40° with a step size of 0.017° and a step time of 50 s.

Differential Scanning Calorimetry. Thermal analysis was carried out using a DSC 131 (Setaram, Diepholz, Germany), which was regularly calibrated using highly-pure standard materials. The DSC measurements were carried out in aluminum crucibles with a constant heating rate of 2 K/min, under a pure helium atmosphere at 8 mL/min.

Fourier Transform Infrared Spectroscopy (FT-IR). A Bruker ALPHA II FT-IR (Bruker, Karlsruhe, Germany) with a diamond attenuated total reflectance (ATR) accessory was used to gather Fourier transform infrared spectra. The solid materials were placed in the ATR cell without further preparation and investigated in the 4000–400 cm^{-1} range, collecting 32 scans at a resolution of 2 cm^{-1}.

Scanning Electron Microscopy. Scanning electron microscopy (SEM) was carried out by using a Carl Zeiss Microscopy Ltd. (Jena, Germany) instrument at an acceleration voltage of 10 keV.

Hot Stage Microscopy. Hot Stage Microscopy was carried out using a Linkam hot stage LTS420 (Linkam Scientific Instruments Ltd., Waterfield, UK) with an Axioskop 2 microscope (Carl Zeiss, Oberkochen, Germany). Images were recorded and analyzed using Axiovision 2 software.

2.5. Solubility Determination

The gravimetric method was used to determine the solubility of AMG and its binary eutectic mixtures in water at room temperature (25 °C). An excess amount of solid eutectic mixture was added to 4 mL water, and the suspension was stirred at 300 rpm at 25 ± 1 °C for 72 h to reach equilibrium. Then it was allowed to settle before liquid phase sampling. The saturated solutions of respective eutectic mixtures were filtered using a 0.45 µm syringe filter and transported in a 5 mL vial. The weight of the full vial with the solution was recorded immediately. The vials were kept in a fume hood for evaporation of the water. Solids were obtained after ~3–4 days. They are further placed in a desiccator for 1 day. All measurements were carried out in duplicate. The solubility of pure AMG (mg/mL) in water was calculated from the product of the mass of the dried material (solubility of eutectic mixture) and the respective AMG content in the dried eutectic mixture (in weight fraction).

3. Results and Discussion

In literature, it was pointed out [1,10] that organic eutectic formation happens when the molecular interaction among identical molecules is comparably stronger than the interaction between different molecules. Till now, there are no exact rules to design eutectics based on the cohesive interactions dominant over the adhesive interaction to produce a eutectic [4,10]. Generally, a cocrystal is expected to be obtained when the possibility of formation of very strong adhesive interactions is high. However, when the cohesive interactions are strong but auxiliary interactions are weak to nil, as long as molecular mismatched shapes are there, then the formation of a eutectic mixture is expected [10]. In the case of AMG, the interactions like imide···imide and N–H···O hydrogen bond interactions are relatively stronger (Figure 1). However, in the case of multicomponent systems with the coformers CAF, NIC and ZMD, the adhesive interactions are relatively weaker as they are non-isomorphous molecules with a considerable mismatch in size and shape. Moreover, the structural arrangement will have lack a unique lattice arrangement distinct from the individual components and thus retains the cohesive interactions in its eutectic mixtures. Therefore, the X-ray diffraction pattern and spectroscopic signature peaks of a eutectic mixture does not contain new crystalline arrangement, rather contains the overlapping of diffraction peaks of the individual components.

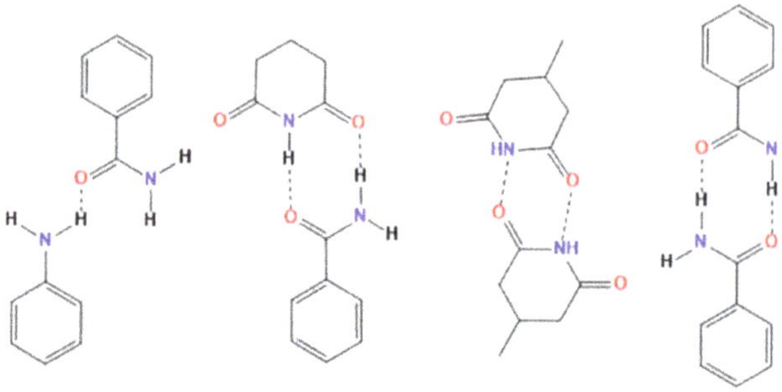

Figure 1. Examples of some probable supramolecular synthons that may be present in AMG-eutectic systems.

The Cambridge Structural Database (CSD) search for imide and carboxylic acid interaction only reveals two reported crystal structures with refcodes HUVGAI and UGOHUV, demonstrating the weaker probability of cocrystal formation via the imide···carboxylic acid heterosynthon. Similarly, no imide and only one monosubstituted pyridine interaction have been observed from CSD search for the multicomponent crystalline systems of imide functional molecules. There is also no reported structure with imide···carboxamide interaction. Furthermore, cocrystal screening by considering the molecular complimentary screening wizard of CSD [24] also reveals no-hit for cocrystals for the selected coformers.

Powder X-ray Diffraction (PXRD) Analysis

X-ray diffraction patterns can be used as one of the confirmatory tools for the detection of the formation of a eutectic system. Figure 2 displays the comparison between the PXRD patterns obtained for starting components and their respective eutectic systems.

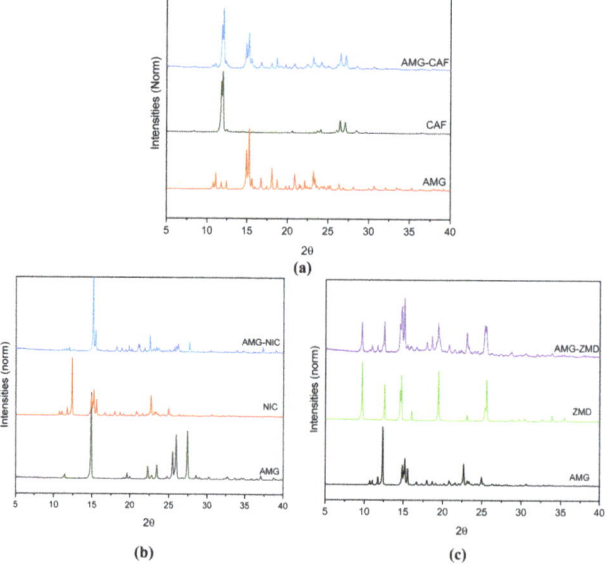

Figure 2. Overlay of PXRD patterns of (**a**) AMG, CAF and the eutectic mixture of the AMG-CAF system, (**b**) AMG, NIC and the eutectic mixture of the AMG-NIC system and (**c**) AMG, ZMD and the eutectic mixture of the AMG-AMD system. Main characteristic peaks are indicated by dashes lines.

In a eutectic system, the retention of the crystal structure of individual components is expected, consequently, the PXRD pattern of the eutectic system should contain the overlap of diffraction peaks of starting components without the generation of new diffraction peaks. As seen in the comparison of the diffraction patterns of the eutectic mixtures and the pure starting components, no new peaks were observed and all the diffraction peaks of the starting components are present in the grinded mixtures, signifying the formation of eutectic mixtures in all three studied cases.

Differential Scanning Calorimetry (DSC) Analysis

DSC is one essential technique to characterize the solid phase behavior of a material and mixtures. It is crucial to detect the eutectic nature of a multicomponent system as typically the melting point of the eutectic mixture is lower than either of the parent components and the eutectic melting can be identified via the melting events occurring. The DSC patterns of the various AMG-coformer mixtures overlayed with those of the pure components are shown in Figure 3, specifically for the AMG-CAF system in Figure 3a, the AMG-NIC system in Figure 3b and the AMG-ZMD system in Figure 3c.

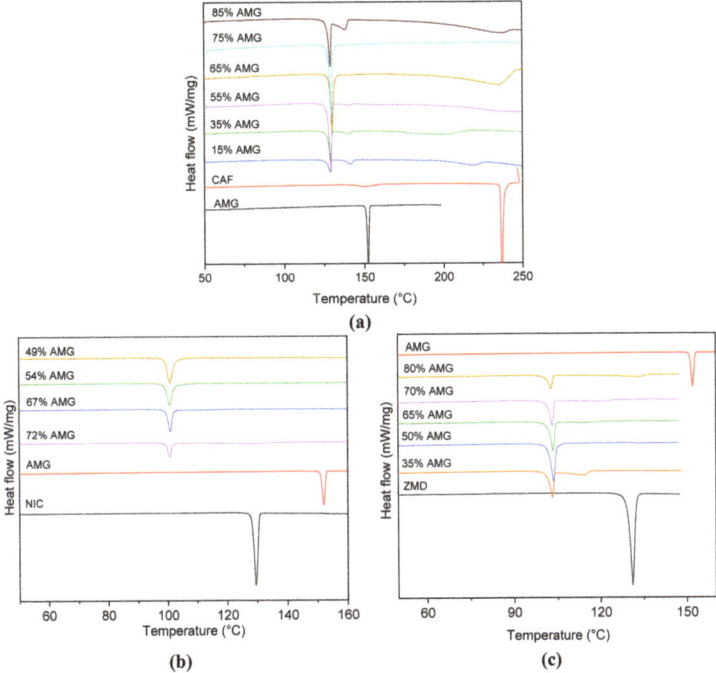

Figure 3. DSC patterns of the different AMG-coformer compositions showing the melting point depressions and the eutectic peak, (a) AMG and CAF, (b) AMG and NIC, and (c) AMG and ZMD.

The thermogram of AMG exhibits only one endothermic melting peak, suggesting no polymorphic transition during the heating cycle. The eutectic melting peak for the AMG-NIC system (Figure 3b) is found at 99.2 °C, being, as expected, significantly lower than for both the starting materials. A close-to single eutectic peak is observed for the ~50% mixture, indicating the presence of the "purely" eutectic phase and thus, the eutectic composition in the AMG-NIC system. In the case of the AMG-CAF system at different weight% ratios, except at the eutectic composition, DSC curves exhibited three endothermic transition effects (see Figure 3a). The first endothermic peak signifies the eutectic temperature, whereas another endotherm signifies the melting of excess AMG or CAF, respectively. The supplementary peak observed at 137.7 °C corresponds to the enantiotropic phase transition of caffeine polymorphic form II to form I [25,26]. The eutectic composition is found to be close to 75 wt.% AMG composition. For the AMG-ZMD system, the melting endothermic peak at 101.8 °C refers to the eutectic melting (see Figure 3c), while the eutectic composition is found close to the AMG-ZMD composition of ~50 wt.%. Table 1 contains the measured temperatures and enthalpies of melting for the pure AMG, the coformers used and their eutectic mixtures. Hot stage microscopic analysis results of the eutectic mixtures which are presented in Figure S1 (Supporting Information), also support the eutectic nature of the AMG mixtures exhibiting a uniform melting in a narrow temperature range.

Table 1. Melting point (M.P) and melting enthalpy (M.E) of drug AMG, the coformers and their eutectic compositions.

Drug	Coformer	M.P (°C)	M.E (J/g)	Eutectic	M.P (°C)	M.E (J/g)
AMG	CAF	235.9	128.51	AMG-CAF	127.2	74.21
M.P = 151.0 (°C)	NIC	128.68	186.81	AMG-NIC	99.2	125.55
M.E = 110.23 J/g	ZMD	129.4	174.7	AMG-ZMD	101.8	124.15

Infrared Spectroscopy

IR spectroscopy was used to understand the probable intermolecular interactions present in the eutectic system. The observed IR spectra of AMG, the coformers and the respective eutectic systems are presented in Figure 4. As seen, the IR spectra of the eutectic systems are comparable with their starting components. No new peak was observed in the three eutectic systems and the spectra simply overlap with the spectra of the individual components. All the characteristic vibrational bands of AMG and the coformers are observed in the spectrum of the eutectic composition. For the AMG spectrum, the N–H starching peak is observed at 3470 cm^{-1}, and the C=O stretch vibration is observed at 1713 cm^{-1}. The slight shifting of C=O peaks in the mixture's absorption bands around 1713–15 cm^{-1} might be caused by the long-range interaction between the carbonyl groups and other functionality in the molecules in the eutectic mixture.

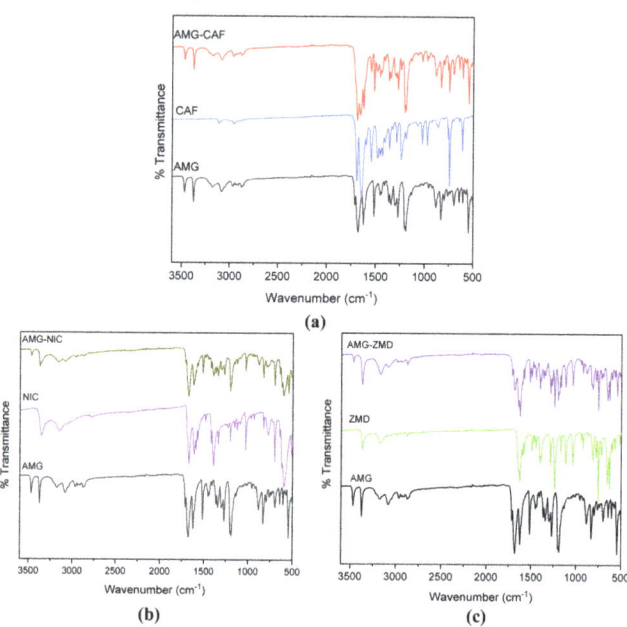

Figure 4. Comparison of FT-IR spectra of starting materials AMG and coformer with their respective eutectic composition for (**a**) AMG and CAF, (**b**) AMG and NIC, and (**c**) AMG and ZMD.

Scanning Electron Microscopy (SEM) Analysis

The micrographs obtained from SEM are shown in Figure 5. Even all samples were grinded in a similar way; SEM images contrast the morphological characteristics of the coformers and partly of the AMG, showing some residual crystals left. However, in the eutectic systems i.e., AMG-CAF, AMG-ZMD and AMG-NIC, a noteworthy reduction in particle size is observed with the appearance of strongly agglomerating homogeneous fines as compared to the parent components.

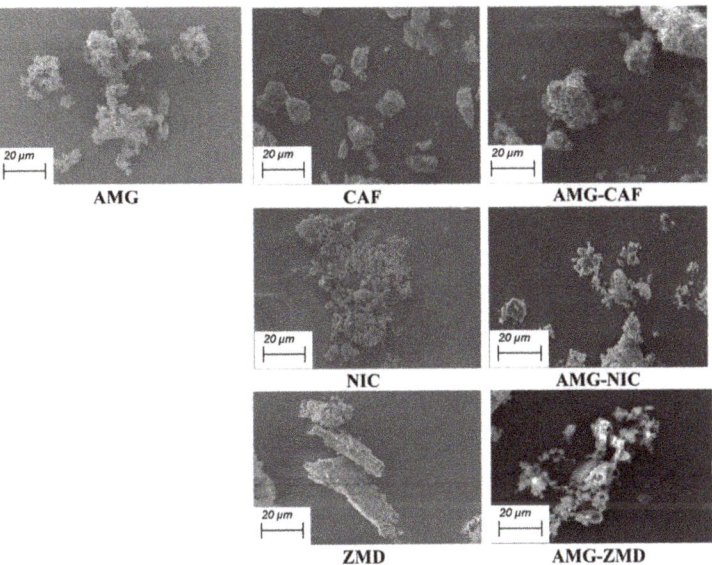

Figure 5. Comparison of micrographs of pure AMG and conformers after LAG treatment with their respective eutectic mixtures at 1.0 Kx magnification.

Phase Diagrams and Related Tammann Plots

From the integration of the eutectic melting effects at different mixture compositions, the related eutectic melting enthalpy ΔH_{eut} can be obtained. At the eutectic composition, the melting enthalpy is maximum and gradually decreases with composition towards the pure components. Therefore, plotting the ΔH_{eut} values as a function of the composition (Tammann plot), the intersection of the two linearized parts provides a good measure of the eutectic composition in the respective binary system.

AMG-CAF system. The presence of the single melting endotherm at 127.2 °C (see Figure 3a), lower than that of both the starting components AMG and CAF, indicated the formation of the eutectic composition in the AMG-CAF system at 75 wt.% of AMG. The phase diagram and the Tammann plot for this system are presented in Figure 6a,b. In the phase diagram, the liquidus and solidus temperatures are plotted as a function of weight% (wt.%) of AMG in the CAF-AMG system. It shows the enantiotropic behavior of caffeine and specifies the polymorphic phase transition at ~138°C confirming the literature data [25,26]. Additionally, this indicates the eutectic composition at ~75 wt.%, which is verified in the Tammann plot (Figure 6b). In addition, the latter indicates a partial solid solution behavior of CAF in AMG in a rather narrow composition range of ~95 ± 2 wt.% of AMG (indicated by the dashed line in Figure 6b). However, since this is based on only two points, it needs to be verified what is hardly to be done in the small existence region of AMG in the phase diagram.

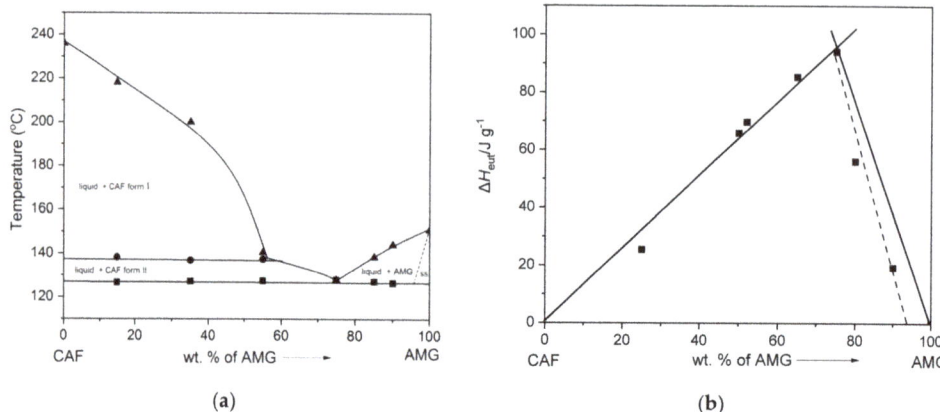

Figure 6. (a) Binary melt phase diagram and (b) Tammann plot of the AMG-CAF system.

AMG-NIC system. The DSC profiles in Figure 3b illustrate the melting behavior of solid samples at different AMG-NIC compositions prepared by neat grinding. The system shows simple eutectic behavior; thus, mixtures of both components exhibit two endothermic peaks in DSC curves corresponding to eutectic melting and the subsequent dissolution effect of the excess compound. Figure 7a presents the derived melt phase diagram with a eutectic composition of ~50 wt.% AMG. The eutectic temperature in this system was found on average at 99.2 °C (Table 1). The Tammann plot shown in Figure 7b confirms the ~50 wt.% eutectic composition.

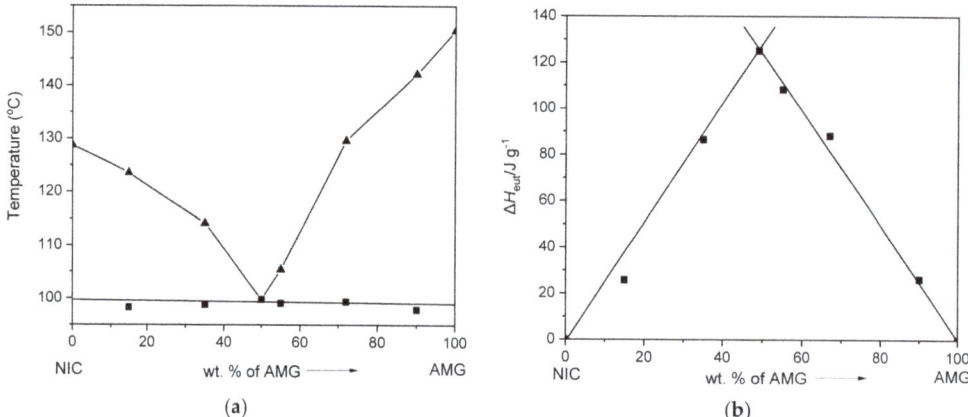

Figure 7. (a) Binary melt phase diagram and (b) Tammann plot of the AMG-NIC system.

AMG-ZMD system. As for the AMG-NIC system, the melt phase diagram of the AMG-ZMD system is characterized by a simple eutectic, possessing a eutectic point at 101.8 °C and composition of ~50 wt.% of AMG. The eutectic composition was derived from the single endothermic event in the DSC thermogram and from the intersection of the ZMD and AMG liquidus curves. Tammann's plot (Figure 8b) for the AMG-ZMD system verifies the eutectic composition close to 50 wt.% AMG and indicates partial miscibility at the solid-state close to the ZMD side (indicated by the dashed line in Figure 8b), but similar to the AMG-CAF system, this needs more detailed studies if important.

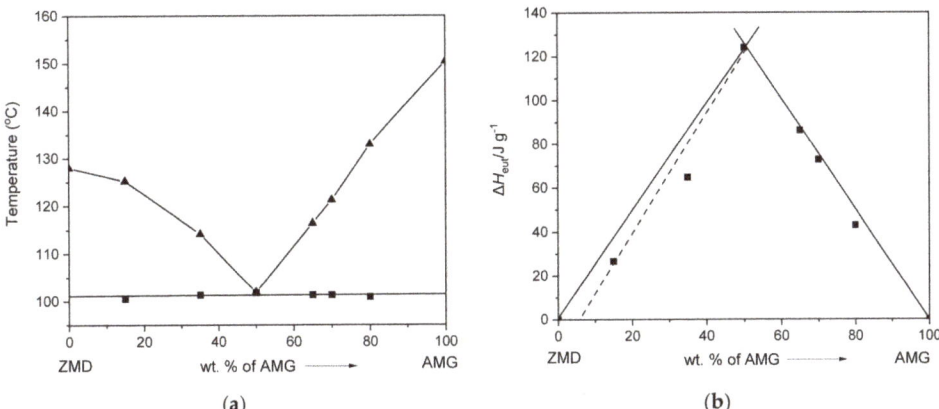

Figure 8. (a) Binary melt phase diagram and (b) Tammann plot of the AMG-ZMD system.

Aqueous Solubility of the Eutectics

To measure the solubility, each experiment was performed in a jacketed, circulating flask maintained at 25 °C and in two sets to ensure consistency. Interestingly, the eutectics show significantly improved solubility over the parent drug AMG with a solubility of 1.9 ± 0.01 mg/mL at 25 °C (See Figure 9). The AMG-CAF, AMG-NIC and AMG-ZMD eutectics exhibit nearly 2.5-fold, 1.5 fold and 1.3 times higher solubility than AMG.

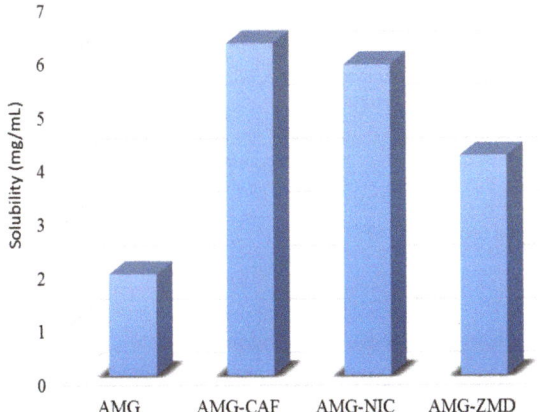

Figure 9. Comparison of solubilities of AMG and its eutectic mixtures in water at 25 °C.

Different thermodynamic parameters like melting temperature depression and the heat of fusion (ΔH_f) can affect the overall solubility [27,28]. Considerable decrease in the ΔH_f (see Table 1) is observed for the AMG-CAF system and expectedly it is showing higher solubility than the AMG. However, higher ΔH_f has been observed for the eutectics AMG-ZMD and AMG-NIC, and as a result, the solubility difference is not significant. Particularly, despite having higher soluble coformer NIC, the eutectic AMG-NIC exhibited less solubility compared to AMG-CAF. The experimental solubility results are supported by the ideal solubility data for the eutectic mixtures calculated by using the van't Hoff equation (Figure S2, Supporting Information), showing similar trends for the solubilities. Moreover, the precipitates obtained from the solubility experiment were further evaluated by PXRD measurements, which are presented in Figure S3 (Supporting Information). The PXRD patterns observed from the eutectics before and after solubility evaluations contain the characteristic peaks for the starting materials and, in addition are similar. Thus, they confirmed no new phase or cocrystal formation during solubility experiments.

4. Conclusions

Solvent-assisted mechanochemical grinding of the drug aminoglutethimide with pharmaceutical coformers enabled the production of binary eutectic mixtures. Considerable lowering of the melting points in the eutectic mixtures was determined from DSC analysis. Analogously, the aqueous solubility of the binary eutectic mixtures was improved compared to the pure AMG. Solubilities at 25 °C were found to be 1.3–2.5 times higher than for the AMG. The final aim of the work was to improve the AMG solubility through physical form alteration, and this objective was achieved by the formation of binary eutectic mixtures with caffeine, nicotinamide and ethenzamide as drug-drug eutectics, with the most significant improvement for the AMG-CAF system.

Supplementary Materials: The following supporting information can be downloaded at: https://www.mdpi.com/article/10.3390/cryst12010040/s1, Figure S1: Comparison of Hot Stage Microscopic images for the eutectic mixtures at room temperature, melting and subsequent cooling. Figure S2: Van't Hoff plot of ideal solubility values for the eutectics., Figure S3: PXRD of the AMG eutectics after solubility determinations.

Author Contributions: Conceptualization, B.S. and H.L.; methodology, B.S.; formal analysis, B.S.; investigation, B.S.; writing—original draft preparation, B.S.; writing—review and editing, B.S. and H.L.; supervision H.L. and A.S.-M. All authors have read and agreed to the published version of the manuscript.

Funding: This research received no external funding.

Institutional Review Board Statement: Not applicable.

Informed Consent Statement: Not applicable.

Data Availability Statement: The data presented in this study are available in manuscript and Supplementary Materials.

Acknowledgments: We thank Stefanie Oberländer and Jacqueline Kaufmann for their help in PXRD and DSC measurements. Markus Ikert is acknowledged for his help in the collection of SEM images.

Conflicts of Interest: The authors declare no conflict of interest.

References

1. Cherukuvada, S.; Nangia, A. Eutectics as improved pharmaceutical materials: Design, properties and characterization. *Chem. Commun.* **2014**, *50*, 906–923. [CrossRef] [PubMed]
2. Saikia, B.; Pathak, D.; Sarma, B. Variable stoichiometry cocrystals: Occurrence and significance. *CrystEngComm* **2021**, *23*, 4583–4606. [CrossRef]
3. Araya-Sibaja, A.M.; Vega-Baudrit, J.R.; Guillén-Girón, T.; Navarro-Hoyos, M.; Cuffini, S.L. Drug Solubility Enhancement through the Preparation of Multicomponent Organic Materials: Eutectics of Lovastatin with Carboxylic Acids. *Pharmaceutics* **2019**, *11*, 112. [CrossRef] [PubMed]
4. Bazzo, G.C.; Pezzini, B.R.; Stulzer, H.K. Eutectic mixtures as an approach to enhance solubility, dissolution rate and oral bioavailability of poorly water-soluble drugs. *Int. J. Pharm.* **2020**, *588*, 119741. [CrossRef]
5. Moore, M.D.; Wildfong, P.L.D. Aqueous Solubility Enhancement Through Engineering of Binary Solid Composites: Pharmaceutical Applications. *J. Pharm. Innov.* **2009**, *4*, 36–49. [CrossRef]
6. Saikia, B.; Bora, P.; Khatioda, R.; Sarma, B. Hydrogen Bond Synthons in the Interplay of Solubility and Membrane Permeability/Diffusion in Variable Stoichiometry Drug Cocrystals. *Cryst. Growth Des.* **2015**, *15*, 5593–5603. [CrossRef]
7. Wünsche, S.; Yuan, L.; Seidel-Morgenstern, A.; Lorenz, H. A Contribution to the Solid State Forms of Bis(demethoxy)curcumin: Co-Crystal Screening and Characterization. *Molecules* **2021**, *26*, 720. [CrossRef]
8. Alhadid, A.; Mokrushina, L.; Minceva, M. Design of Deep Eutectic Systems: A Simple Approach for Preselecting Eutectic Mixture Constituents. *Molecules* **2020**, *25*, 1077. [CrossRef]
9. Gala, U.; Chuong, M.C.; Varanasi, R.; Chauhan, H. Characterization and Comparison of Lidocaine-Tetracaine and Lidocaine-Camphor Eutectic Mixtures Based on Their Crystallization and Hydrogen-Bonding Abilities. *AAPS PharmSciTech* **2015**, *16*, 528–536. [CrossRef]

10. Cherukuvada, S.; Guru Row, T.N. Comprehending the Formation of Eutectics and Cocrystals in Terms of Design and Their Structural Interrelationships. *Cryst. Growth Des.* **2014**, *14*, 4187–4198. [CrossRef]
11. Boeri, L.; Pozzi, E.; Fallara, G.; Montorsi, F.; Salonia, A. Real-life use of the eutectic mixture lidocaine/prilocaine spray in men with premature ejaculation. *Int. J. Impot. Res.* **2021**. [CrossRef] [PubMed]
12. Hyun, S.-M.; Lee, B.J.; Abuzar, S.M.; Lee, S.; Joo, Y.; Hong, S.-H.; Kang, H.; Kwon, K.-A.; Velaga, S.; Hwang, S.-J. Preparation, characterization, and evaluation of celecoxib eutectic mixtures with adipic acid/saccharin for improvement of wettability and dissolution rate. *Int. J. Pharm.* **2019**, *554*, 61–71. [CrossRef] [PubMed]
13. Vasconcelos, T.; Sarmento, B.; Costa, P. Solid dispersions as strategy to improve oral bioavailability of poor water soluble drugs. *Drug Discov.* **2007**, *12*, 1068–1075. [CrossRef]
14. Goud, N.R.; Suresh, K.; Sanphui, P.; Nangia, A. Fast dissolving eutectic compositions of curcumin. *Int. J. Pharm.* **2012**, *439*, 63–72. [CrossRef] [PubMed]
15. Bi, M.; Hwang, S.-J.; Morris, K.R. Mechanism of eutectic formation upon compaction and its effects on tablet properties. *Thermochim. Acta* **2003**, *404*, 213–226. [CrossRef]
16. Zalac, S.; Khan, M.Z.I.; Gabelica, V.; Tudja, M.; Mestrovic, E.; Romih, M. Paracetamol-Propyphenazone Interaction and Formulation Difficulties Associated with Eutectic Formation in Combination Solid Dosage Forms. *Chem. Pharm. Bull.* **1999**, *47*, 302–307. [CrossRef] [PubMed]
17. Harris, A.L.; Dowsett, M.; Stuart-Harris, R.; Smith, I.E. Role of aminoglutethimide in male breast cancer. *Br. J. Cancer* **1987**, *54*, 657–660. [CrossRef]
18. Santen, R.J.; Misbin, R.I. Aminoglutethimide: Review of Pharmacology and Clinical Use. *Pharmacother. J. Hum. Pharmacol. Drug Ther.* **1981**, *1*, 95–119. [CrossRef] [PubMed]
19. Manni, A. Endocrine therapy of metastatic breast cancer. *J. Endocrinol. Investig.* **1989**, *12*, 357–372. [CrossRef]
20. Dahan, A.; Miller, J.M.; Amidon, G.L. Prediction of solubility and permeability class membership: Provisional BCS classification of the world's top oral drugs. *AAPS J.* **2009**, *11*, 740–746. [CrossRef]
21. Faudone, G.; Arifi, S.; Merk, D. The Medicinal Chemistry of Caffeine. *J. Med. Chem.* **2021**, *64*, 7156–7178. [CrossRef] [PubMed]
22. Gehring, W. Nicotinic acid/niacinamide and the skin. *J. Cosmet. Dermatol.* **2004**, *3*, 88–93. [CrossRef]
23. Aitipamula, S.; Chow, P.S.; Tan, R.B.H. Trimorphs of a pharmaceutical cocrystal involving two active pharmaceutical ingredients: Potential relevance to combination drugs. *CrystEngComm* **2009**, *11*, 1823–1827. [CrossRef]
24. Macrae, C.F.; Sovago, I.; Cottrell, S.J.; Galek, P.T.A.; McCabe, P.; Pidcock, E.; Platings, M.; Shields, G.P.; Stevens, J.S.; Towler, M.; et al. Mercury 4.0: From visualization to analysis, design and prediction. *J. Appl. Crystallogr.* **2020**, *53*, 226–235. [CrossRef] [PubMed]
25. Dichi, E.; Sghaier, M.; Guiblin, N. Reinvestigation of the paracetamol–caffeine, aspirin–caffeine, and paracetamol–aspirin phase equilibria diagrams. *J. Therm. Anal. Calorim.* **2018**, *131*, 2141–2155. [CrossRef]
26. Dichi, E.; Legendre, B.; Sghaier, M. Physico-chemical characterisation of a new polymorph of caffeine. *J. Therm. Anal. Calorim.* **2014**, *115*, 1551–1561. [CrossRef]
27. Lorenz, H. Solubility and Solution Equilibria in Crystallization. In *Crystallization: Basic Concepts and Industrial Applications*; Beckmann, D.W., Ed.; Wiley-VCH: Weinheim, Germany, 2013; pp. 35–74.
28. Tao, M.; Wang, Z.; Gong, J.; Hao, H.; Wang, J. Determination of the Solubility, Dissolution Enthalpy, and Entropy of Pioglitazone Hydrochloride (Form II) in Different Pure Solvents. *Ind. Eng. Chem. Res.* **2013**, *52*, 3036–3041. [CrossRef]

Article

Oxalic Acid, a Versatile Coformer for Multicomponent Forms with 9-Ethyladenine

Mónica Benito [1,*], Miquel Barceló-Oliver [2], Antonio Frontera [2] and Elies Molins [1,*]

1 Institut de Ciència de Materials de Barcelona (ICMAB-CSIC), Campus de la Universitat Autònoma de Barcelona, 08193 Bellaterra, Spain
2 Departament de Química, Universitat de les Illes Balears, Ctra. Valldemossa km 7.5, 07122 Palma de Mallorca, Spain; miquel.barcelo@uib.es (M.B.-O.); toni.frontera@uib.es (A.F.)
* Correspondence: mbenito@icmab.es (M.B.); elies.molins@icmab.es (E.M.)

Abstract: Six new multicomponent solids of 9-ethyladenine and oxalic acid have been detected and characterized. The salt screening has been performed by mechanochemical and solvent crystallization processes. Single crystals of the anhydrous salts in 1:1 and 2:1 nucleobase:coformer molar ratio were obtained by solution crystallization and elucidated by single-crystal X-ray analysis. The supramolecular interactions observed in these solids have been studied using density functional theory (DFT) calculations and characterized by the quantum theory of "atoms in molecules" (QTAIM) and the noncovalent interaction plot (NCIPlot) index methods. The energies of the H-bonding networks observed in the solid state of the anhydrous salts in 1:1 and 2:1 nucleobase:coformer are reported, disclosing the strong nature of the charge assisted NH···O hydrogen bonds and also the relative importance of ancillary C–H··O H-bonds.

Keywords: nucleobases; multicomponent solids; crystal engineering; DFT; H-bonding

1. Introduction

The preparation of alternative salts, and more recently cocrystals, of many active pharmaceutical ingredients (APIs) has gained great attention as an important strategy in crystal engineering for the last decades. The use of salts and cocrystals for improving the physico-chemical properties of APIs has been demonstrated, as these solids can modify the solubility or dissolution rate, or even improve the physical stability. Around a half of the marketed products are estimated to be sold as salt forms [1]. Oxalic acid is among the different salt-forming acids listed for this purpose. It belongs to the second class of salt formers according to the *Handbook of Pharmaceutical Salts: Properties, Selection and Use* [1], as they are not naturally occurring but during their application have shown low toxicity and good tolerability. Through a search in the FDA Orange Book Database, we found only two approved examples, which are the specialty dosage forms commercialized as Lexapro® and Movantik®. The first one, in the market since 2009, is used for the treatment of depression and anxiety. Interestingly, it is not a classical salt, but in fact it is a hydrated cocrystal of a salt, composed of protonated escitalopram cations, oxalate dianions, unprotonated oxalic acid molecules and water molecules [2]. The second dosage form contains naloxegol oxalate, indicated for the treatment of opioid-induced constipation in adults with chronic non-cancer pain. Curiously, it was found through a small-scale screening due to the difficulties in preparing naloxegol in solid form [3]. As a result, two polymorphic forms were described, and the crystal structure resolution of Form B showed that it contained hydrogen oxalate anions [4]. In spite of the low incidence of oxalic acid in the list of the most used anions in APIs [5], many other examples have also been described in the literature exhibiting the interest of the scientific community for this coformer [6–9].

Among pharmaceutical compounds, nucleobases are of great value thanks to their contribution as structural components of several pharmaceuticals as well as for their biological function as part of DNA and RNA. Very recently, a new cytosine derivative has been announced for COVID-19 oral treatment, molnupiravir [10].

After a careful revision of cocrystals or salts between oxalic acid and nucleobases or related compounds only a few examples resulted. These include pure adenine [11], N[6]-benzyladenine [12], or the following xanthines: caffeine [13,14], theobromine [15] and theophylline [16,17]. The only example described as a cytosine derivative was for the antiretroviral compound, lamivudine [9].

9-Ethyladenine (9ETADE) is a modified nucleobase with the amino group at N(9) blocked by an ethyl chain (Scheme 1a). We have previously reported the ability of this modified nucleobase to form different salts and/or cocrystals with alkyl dicarboxylic acids (HOOC-X$_n$-COOH, with n from 1 to 4) as coformers depending on the ΔpK$_a$ [18]. Now, as a continuation of our work related to this modified purine, we describe herein the special situation with oxalic acid. This is the most acidic (pK$_a$ = 1.19) and shortest alkyl dicarboxylic acid we have studied but the most versatile and promiscuous in rendering several solid forms.

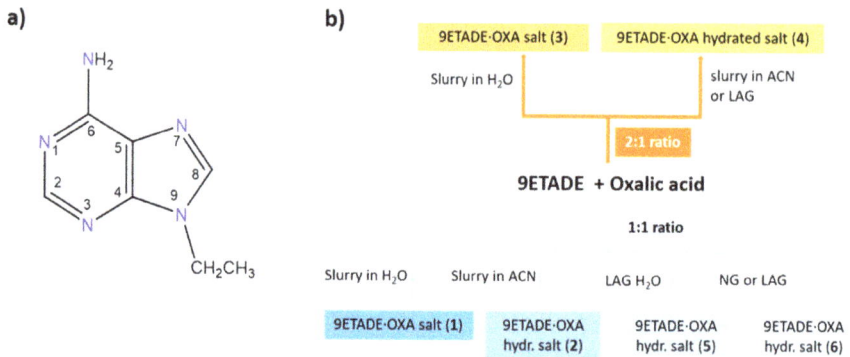

Scheme 1. (a) Molecular structure of 9-ethyladenine with numbering and (b) compounds of 9-ethyladenine with oxalic acid prepared in this work.

Herein, we present several multicomponent forms obtained by combination of the model compound 9-ethyladenine and oxalic acid, which have been prepared by solvent/slurry crystallization or mechanochemistry (liquid assisted grinding (LAG) or neat grinding (NG)) and their full characterization. The crystal structures of the two anhydrous salts have been solved. Moreover, density functional theory (DFT) calculations were used to understand/study their supramolecular interactions focusing on the energetics of the H-bonds, which were computed by carrying out a topological analysis of the electron density.

2. Materials and Methods

2.1. Materials

All reagents were purchased from Sigma-Aldrich Co. (Merck KGaA, Darmstadt, Germany) and used without further purification. Analytical grade solvents were used for the crystallization experiments. 9-Ethyladenine (9ETADE) was obtained as previously described by some of us [19].

2.2. Syntheses of Multicomponent Solids

Solution syntheses. In general, mixtures of 9ETADE and oxalic acid dihydrate in 1:1 or 2:1 molar ratio were suspended in Milli-Q water (for **1** and **3**) or in acetonitrile (for **2** and **4**) at room temperature for two days. After that, the solids were filtered and air-dried. Additional detailed data are included in Supplementary Information (SI).

Grinding Screening. As a general method, mechanochemical syntheses of the new compounds were performed using a Retsch MM400 mixer mill (Retsch, Haan, Germany) in 10 mL agate grinding jars with two 5 mm agate balls. Mixtures containing 1:1 or 2:1 stoichiometric molar ratio of 9ETADE and oxalic acid dihydrate and the selected solvent were ground for 30 min at 30 Hz. Additional details are included in Supplementary Information (SI).

2.3. Characterization

Powder X-ray Diffraction (PXRD). PXRD data were collected using a Siemens D5000 powder X-ray diffractometer (Siemens, Munich, Germany) with Cu-Kα radiation (λ = 1.5418 Å), with 35 kV and 45 mA voltage and current applied. An amount of powder was gently pressed on a glass slide to afford a flat surface and then analyzed. The samples were scanned in the 2θ range of 2–50° using a step size of 0.02° and a scan rate of 1 s/step.

Single Crystal X-ray Diffraction (SC-XRD). Suitable crystals of **1** and **3** were selected for X-ray single crystal diffraction experiments, covered with oil (Infineum V8512, formerly known as Paratone N) and mounted at the tip of a nylon CryoLoop on a BRUKER-NONIUS X8 APEX-II KAPPA CCD diffractometer (Bruker, Karlsruhe, Germany) using graphite monochromated MoKα radiation (λ = 0.7107 Å). Crystallographic data were collected at 300 (2) K. Data were corrected for Lorentz and polarization effects and for absorption by SADABS [20]. The structural resolution procedure was made using the WinGX package [21]. The structure factor phases were solved by SHELXT-2014/5 or SHELXT-2018/2 [22]. For the full matrix refinement SHELXL-2017/1 or SHELXL2018/3 was used [23]. The structures were checked for higher symmetry with help of the program PLATON [24]. H-atoms were introduced in calculated positions and refined riding on their parent atoms, except for the protonation sites (H1A and H1B).

In Table 1 general and crystallographic data for the two new salts described are summarized.

Table 1. Crystallographic data and refinement for salts **1** and **3**.

Crystal	1	3
Empirical Formula	$C_9H_{11}N_5O_4$	$C_8H_{10}N_5O_2$
Mr	253.23	208.21
Crystal system	Triclinic	Monoclinic
Space group	$P\bar{1}$	$P2_1/c$
a/Å	5.5478 (19)	10.325 (4)
b/Å	9.413 (3)	7.0783 (3)
c/Å	11.535 (4)	13.207 (5)
α/°	70.272 (5)	90
β/°	87.851 (5)	106.457 (5)
γ/°	81.231 (6)	90
V/Å^3	560.3 (3)	925.6 (6)
Z	2	4
Radiation type	Mo K$_\alpha$	Mo K$_\alpha$
μ/mm^{-1}	0.121	0.113
Temperature/K	300 (2)	300 (2)
Crystal size/mm	0.500 × 0.230 × 0.180	0.390 × 0.120 × 0.080
D_{calc}/g·cm^{-3}	1.501	1.494
Reflections collected	6719	1676
Independent Reflections	2581 [R(int) = 0.0277]	1676 [R(int) = 0.062]
Completeness to theta = 24.996°	99.8%	99.2%
F(000)	264	436
Data/restraints/parameters	2581/0/170	1676/0/140
Goodness-of-fit	1.056	1.090
Final R indices [I < 2d(I)]	R1 = 0.0420, wR2 = 0.1147	R1 = 0.0610, wR2 = 0.1465
R indices (all data)	R1 = 0.0478, wR2 = 0.1192	R1 = 0.0863, wR2 = 0.1540
Largest diff. peak and hole/e·Å^{-3}	0.318 and −0.222	0.235 and −0.275
CCDC n°	2126070	2126069

Thermogravimetric analysis—Differential scanning calorimetry (TGA-DSC). A simultaneous thermogravimetric analysis (TGA)—differential scanning calorimetry/differential thermal analysis (heat flow DSC/DTA) system NETZSCH -STA 449 F1 Jupiter (NETZSCH, Selb, Germany) was used to perform thermal analysis on the solids. Samples (3–8 mg) were placed in open alumina pan and measured at a scan speed of 10 $°C\ min^{-1}$ from ambient temperature to 250 °C under N_2 atmosphere as protective and purge gas (their respective flow velocities were 20 and 40 mL/min).

Attenuated Total Reflection Fourier Transform Infrared spectroscopy (ATR-FT-IR). A Jasco 4700LE spectrophotometer (JASCO, Tokyo, Japan) with attenuated total reflectance accessory was used to record the FT-IR spectra of 9ETADE, oxalic acid dihydrate and the new compounds prepared in this work in the range from 4000 to 400 cm^{-1} and at a resolution of 4.0 cm^{-1}.

Determination of approximate solubilities. The approximate solubilities of the anhydrous 9ETADE salts were determined by the gravimetric method following the procedure described in [25,26]. To sum up, in a vial, an amount of solid (ca. 40–50 mg) was added a determined volume of Milli-Q water to obtain a supersaturated solution at room temperature. The suspensions were stirred for 2 h and then the agitation was stopped to allow slow settling of the solids in excess for at least 24 h. Samples of the supernatant liquid were taken using a syringe and filter via a nylon syringe filter (0.22 μm). The clear solutions were added to a pre-weighted vial (m_1) and the vial was weighted again (m_2). The solvent was allowed to evaporate in the fume hood until dry and the mass was recorded (m_3). The solids were dried in an oven at 30 °C under vacuum for 2 h to confirm no further weight decrease. The solubility was calculated as the amount of solid recovered ($m_3 - m_1$) divided by the volume of the solution ($m_2 - m_3$). The given values are the median of three replicates. The residual solids which did not dissolve were analyzed by PXRD to check the stability of the salts.

2.4. Stability Studies

Stability studies in solution. Mixtures of 9ETADE and oxalic acid in 1:1 ratio and using Milli-Q water or acetonitrile as a solvent were conducted at room temperature at different times (2, 24 and 48 h) to follow phase evolution.

Physical stability in solid state. The physical stability of the two anhydrous salts was carried out by exposing the solids to the certain range of relative humidity (RH). The humidity inside the chamber/desiccator was maintained at 75% RH using a sodium chloride saturated salt solution [27]. The samples were exposed for one week at 40 °C and 1 month at room temperature and subsequentially analyzed by PXRD.

2.5. Theoretical Calculations

Gaussian-16 [28] program package was used for the calculations reported in this work. For the energies and wavefunction calculations we have used the PBE0 [29] functional in combination with Grimme's dispersion correction (D3) [30] and the triple-ζ basis set def2-TZVP [31]. The molecular electrostatic potential (MEP) surfaces have been computed at the same level of theory and using the 0.001 a.u. isosurface to emulate the van der Waals surface. The experimental crystallographic coordinates have been used to evaluate the interactions in the solid state. This level of theory and methodology was used before to investigate similar interactions [32–35]. The Bader's quantum theory of "Atoms in molecules" (QTAIM) [36] and the noncovalent interaction index (NCIPlot) index [37] methods were used to characterize the hydrogen bonding interactions by means of the AIMall program [38]. The dissociation energies of the H-bonding interactions have been evaluated using the methodology proposed by Espinosa et al. [39] based on the potential energy density (Vr) at the bond critical points (CPs).

3. Results and Discussion

3.1. Solid-State Characterization of the Solids Obtained from Slurry or Mechanochemical Methods

The cocrystallization of 9-ethyladenine and oxalic acid was carried out in 1:1 or 2:1 (9ETADE:OXA) molar ratios. On the one hand, by mechanochemical process by neat grinding (NG) or liquid-assisted grinding (LAG), which have previously demonstrated to be fast methods for salt/cocrystal screenings. On the other hand, slurry experiments in water or acetonitrile were also performed with the purpose to check whether other plausible solid forms or solvates were possible. In solution, we found that for these two molar ratios and the two solvents, up to four different powder patterns were obtained (see Figure 1). The solids were identified as shown in Scheme 1b (compounds **1–4**).

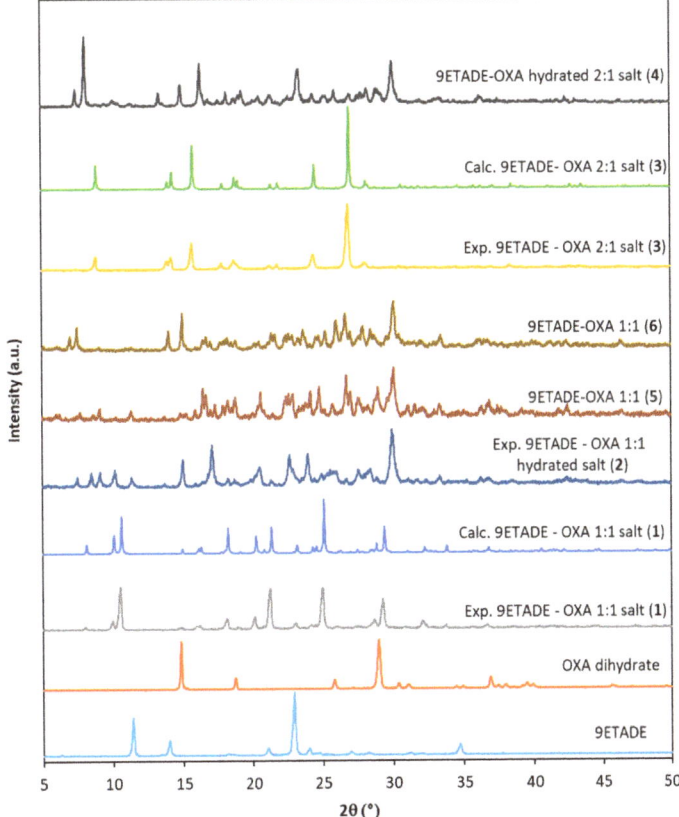

Figure 1. Comparison of the experimental and calculated powder patterns of 9ETADE, oxalic acid dihydrate and the as-synthesized compounds **1–6**.

During the mechanochemical screening, for the 2:1 molar ratio by NG or LAG in methanol or water, the powder patterns of the resulting solids were identical to the diffractogram of compound **4**. However, for the 1:1 ratio, the situation was much more complex. Use of water or methanol afforded powder patterns not only different between them but also to the ones obtained in solution with water or acetonitrile. The solid obtained by LAG in water was identified as compound **5** and the solids obtained by NG or by LAG in methanol, as compound **6**. The diffractograms of solids **5** and **6** are shown in Figure 1).

Between the powder patterns of solids **2** and **5** some differences can be observed, and therefore they will be treated as two different forms. Even if these two solid forms show some characteristics in common not only in their FT-IR spectra but also in their TGA-DSC traces, as it will be commented below. Among all the other forms or respect to the starting products, characteristic and distinguished peaks are clearly observed. Furthermore, the agreement among the experimental and calculated patterns from single crystal data for compounds **1** and **3** was excellent. Surprisingly, the shortest dicarboxylic acid we used resulted the most fruitful in rendering different solid forms.

All the new compounds were analyzed by Fourier Transform Infrared spectroscopy in Attenuated Total Reflectance mode (ATR-FT-IR) and thermal methods (TGA-DSC) to elucidate the nature of these materials. Changes in the IR spectra of the new solids respect to the starting compounds have provided evidence that new H-bond interactions have taken place. Moreover, the presence or absence of several characteristic vibration modes, as for instance NH_2, COOH, C=N, gave further information of these new contacts. The IR spectra are included in Figure 2.

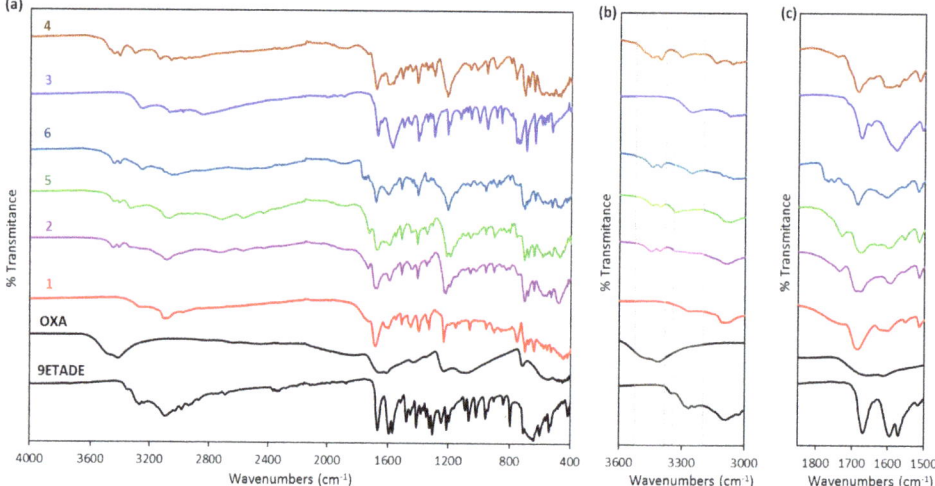

Figure 2. FT-IR spectra of (**a**) 9ETADE, oxalic acid dihydrate and the as-synthesized compounds in the full range, and in the ranges (**b**) 3600–3000 cm^{-1} or (**c**) 1850–1500 cm^{-1}.

The main difference among compounds prepared from aqueous slurries (**1** and **3**) with respect to all the others is the absence of one or two bands in the region 3450–3400 cm^{-1} (see Figure 2). While for these two solids no bands are observed, confirming the anhydrous state observed in the thermal analysis, for all the other compounds (**2**, **4**–**6**), two bands at ca. 3446 and 3407 cm^{-1} were present, which could be assigned to OH from water, and from the hydroxyl of unprotonated oxalic acid. Some differences in the NH region can be also detected.

In the spectra of the starting compounds, the also characteristic ν(C=O), ν(C=N), and δ_{sciss} (NH_2) vibrations appear at 1654, 1610 and 1669 cm^{-1}. For some of the compounds (**1**, **4** and **6**), only a broad band coincides at approximately at 1686 cm^{-1}, or for **3** and **5**, at 1678 cm^{-1}, in agreement with previous observations [9]. However, for the solid **2** a doublet overlapped band at 1694 and 1678 cm^{-1} can be detected. Moreover, for compounds **1**, **2** and **4**–**6**, a small band appears between 1728–1770 cm^{-1}, which confirms again the presence of C=O from unprotonated –COOH from oxalic acid [40]. No band is observed in this region for compound **3**.

By simultaneous thermogravimetric analysis (TGA)—differential scanning calorimetry/differential thermal analysis (DSC/DTA)—the stability of the new salts was evaluated. Both compounds **1** and **3** were anhydrous according to their TGA traces (Figures S1 and S2, respectively). In the DSC trace for compound **3** only one big endothermic peak corresponding to the melting process (melting peak value of 227.5 °C) and concomitant degradation was observed. This is the highest melting point for the solid forms prepared in this work. Thus, this is the most stable form. For compound **1**, the DSC showed a previous small *endo* with an onset temperature of 176.2 °C followed by two overlapped endothermic processes with an onset temperature for the first one at 211.6 °C. The second endothermic peak (T_{peak}, 227.8 °C) agrees with the one observed for compound **3**. This event would suggest the loss of an oxalic acid molecule to achieve the salt **3** in 2:1 molar ratio.

On the contrary, compounds **2** and **4** showed each one a loss on drying (lod), of 2.74 and 4.19%, respectively, until 125 °C (Figures S3 and S4). They were assigned to dehydration processes of 1/3 to 1/2 of molecule of water for **2** (being probably a non-stoichiometric hydrated) and one molecule of water for **4** (theor. lod of one H_2O, 4.16%). In their DSC after these desolvation processes (at T_{peak} 107.2 °C for **2** and T_{peak}, 111.5 °C for **4**), the melting of the solids followed by degradation was observed.

The TGA-DSC trace for compound **5** resembles to the one of **2**. In its TGA a loss of 2.6% was observed closer to the theoretical value for a third hydrate compound (2.32%) and appeared in the same region. Besides, in the DSC an endothermic desolvation process was observed (peak temperature at 111.7 °C) and a subsequent melting (melting peak temperature, 187.9 °C) and its degradation (Figure S5).

Finally, for compound **6**, in the TGA trace a mass loss of about 5.8% was observed before complete degradation of the solid, suggesting that a new solvate was formed. The theoretical loss for a monohydrate solid form is 6.6%. The DSC thermogram showed several endothermic events, unveiling a complex desolvation process for this solid form (Figure S6).

3.2. Single-Crystal Structures

The crystal structures of compounds **1** and **3** were solved by single crystal X-ray diffraction. All trials performed to obtain suitable single crystals for the other multicomponent solids observed in this work were unfruitful. ORTEP images of both compounds are included in Figure S7.

9-Ethyladenine—oxalic acid (1:1) salt (**1**). Crystal structure analysis of compound **1** reveals that this compound crystallizes in the triclinic *P*-1 space group and the asymmetric unit consists of a protonated molecule of 9ETADE and a hydrogen oxalate anion. No solvent molecules are observed in agreement with its FTIR spectrum or the TGA trace where no loss on drying was observed before melting. Moreover, the agreement between the experimental and the simulated patterns for the 9ETADE-oxalic acid (1:1) salt is excellent indicating the high purity of the bulk solid obtained by slurry, Figure 1.

The structure consists of the self-assembly of two adenine moieties through its Hoogsteen edge by N(6)–H$\cdots$N(7) interactions (distance: 2.9469(18) Å; angle 161.6°) forming a $R_2^2(10)$ motif (see Figure 3). Oxalate anions are connected head-to-tail through a strong hydrogen bond (distance: O(1B)–H(1B)$\cdots$O(3B), 2.4728(15) Å). Furthermore, two molecules of 9ETADE interact with two hydrogen oxalate molecules through the NH$_2$ group and N(1) to the carboxylic acids of different oxalate anions, as follows: N(6A)–H$\cdots$O(2B) (distance 2.9438(17) Å; angle 139.9°), N(6A)–H$\cdots$O(3B) (distance 3.0271(18) Å; angle 128.7°), N(1A)–H$\cdots$O(1B) (distance 3.0412(17) Å; angle 124.0(16)°) and N(1A)–H$\cdots$O(4B) (distance 2.7449(16) Å; angle 158.7(18)°) following ring motifs $R_1^2(5)$, and a further C(2A)–H(2A)$\cdots$O(1B) interaction.

The whole structure is formed by the connection of the layers formed by 9ETADE dimers and hydrogen oxalate ions (Figure 3c) through the hydrogen bonds, as follows: C(10A)–H(10A)$\cdots$O(4B) and C(8A)–H(8A)$\cdots$O(2A). The complete list of H-bonding interactions is shown in Table S1.

9-Ethyladenine—oxalic acid (2:1) salt (**3**). Single crystals of this salt were obtained by slow evaporation of a mixture in acetonitrile—water (1:1 vol/vol). In this case, 9ETADE and oxalic acid crystallize in the monoclinic $P2_1/c$ space group containing two protonated molecules of 9ETADE and an oxalate anion in the asymmetric unit forming a salt in a 2:1

ratio. No solvent molecules are observed, again in agreement with the TGA trace or the FTIR spectrum.

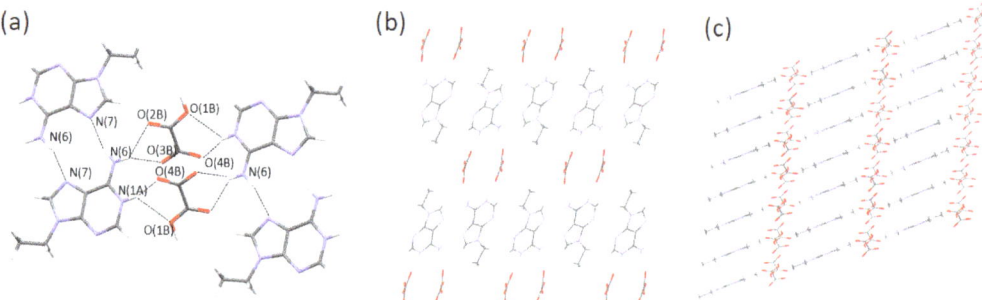

Figure 3. (**a**) Hydrogen bonding interactions, (**b**) layers (view along *a* axis) and (**c**) view along the vector (−2 1 0) in compound **1** (9ETADE-OXA (1:1) salt).

Two molecules of 9ETADE are bridged by an oxalate molecule through the following interactions (Figure 4): N(6A)–H···O(2B) (distance 2.743(3) Å; angle 152.7°), N(1A)–H···O(1B) (distance 2.746(3) Å; angle 148(3)°) and N(1A)–H···O(2B) (distance 2.968(3) Å; angle 133(3)°). An additional hydrogen bond to a second oxalate anion is established through N(6A)–H···O(1B) (distance 2.873(3) Å; angle 156.4°), thus constituting belts perpendicular to the **c** axis. These belts are pilled by hydrogen bonds through the methyl group from the ethyl chain to the carbonyl from the oxalate anion (distance C(11A)–H(11A)···O(2B), 3.576(5) Å). Among 9-ethyladenine molecules, only C–H···N contacts are observed (C(2A)–H(2A)···N(7A), 3.224(4) Å). Surprisingly, the N(7A)–Cg distance (3.398 Å) is shorter than the distance between mean planes for two pilled 9ETADE molecules, suggesting an attractive lone pair–π interaction (N(7A)···6-membered ring). The complete list of hydrogen bonds is included in the Supplementary Information, Table S2.

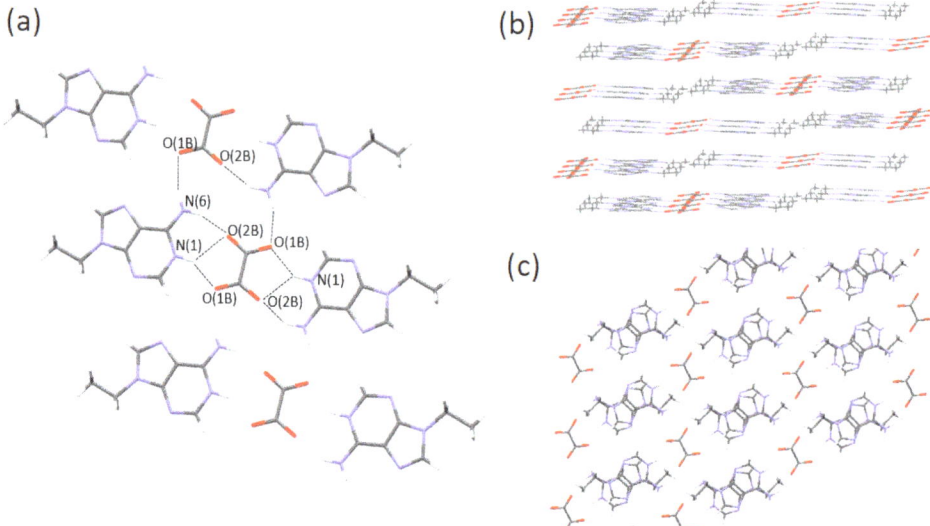

Figure 4. (**a**) Hydrogen bonds and (**b**) packing of the belts (view along the *b* axis, slightly tilted to improve the perspective) and (**c**) π–π stacking (view along the *c* axis) in 9ETADE-oxalic acid (2:1) salt (**3**).

The agreement between the experimental and the simulated powder diffraction patterns for the 9ETADE-oxalic acid (2:1) salt is excellent, Figure 1.

Although the crystal structures for the other multicomponent solids shown in this work were unavailable, the data collected from IR spectroscopy, thermogravimetric analyses, and the knowledge from previous crystal structures regarding to other salts and cocrystals described for this compound allow to determine their composition. So, we hypothesize that compound **4** contains a half double charge oxalate anion and a half unprotonated oxalic acid molecule per 9ETADE molecule. Additionally, for compounds **2**, **5** and **6**, a half mono-oxalate anion and a half unprotonated oxalic acid should be present, although with different contents of water molecules in each case.

3.3. Aqueous Solubilities

One of the main objectives of conducting salt/cocrystal screenings of active molecules is to find suitable candidates with the best physical-chemical properties. In a previous work, we described two new hydrated salts (malonate and fumarate) and an anhydrous succinate of 9ETADE [18]. Herein, the aqueous solubilities of 9-ethyladenine and the three anhydrous candidates (the two anhydrous oxalate salts prepared in this work and the succinate salt of 9ETADE) were determined for comparative purposes (Table 2).

Table 2. Aqueous solubilities of the succinate salt, compounds **1** and **3** and the precursors.

Compound	Solubility (mg/mL)	Coformer	Solubility * (mg/mL)
9ETADE	533		
9ETADE-OXA (1:1) salt (**1**)	25	Oxalic acid dihydrate	100
9ETADE-OXA (2:1) salt (**3**)	10		
9ETADE-SUC (1:1) salt	7	Succinic acid	83

* Sigma-Aldrich.

From these results, the modified nucleobase showed the highest solubility in water in comparison to the other salts or even the former carboxylic acids used. Although the results did not indicate an improvement in the solubility behavior of the new compounds, it shows that it is possible to modulate this property, resulting apparently the 9ETADE-OXA (1:1) solid form the most soluble of the three studied salts.

The undissolved residues were analyzed by PXRD. It is worth noting that the compound **1** had transformed partially to the 2:1 salt, compound **3**. Thus, this case is a new example of a solvent-mediated phase transformation and the solubility value obtained for this compound should be discarded. Nevertheless, the results obtained indicated that the powder patterns for this compound **3** and the succinate (1:1) salt remained undisturbed (Figure S8).

3.4. Stability Studies in Solution and Solid State

In order to understand the phase stability of the multicomponent forms in 1:1 molar ratio in solution, the process was investigated over time by PXRD. After 2 h, either using water or acetonitrile resulted to be almost undistinguished products, as the XRD patterns were quite similar. However, the slurry in water after 24 h contained a mixture of the previous phase and a new one, which resulted the unique after 48 h and matched with the diffractogram obtained for compound **1** (Figure 5, below). For the mixture in acetonitrile, the phase transformation was also detected by the presence of new peaks in the powder pattern after stirring for 24 h, which after 48 h resulted in the same PXRD as for compound **2**.

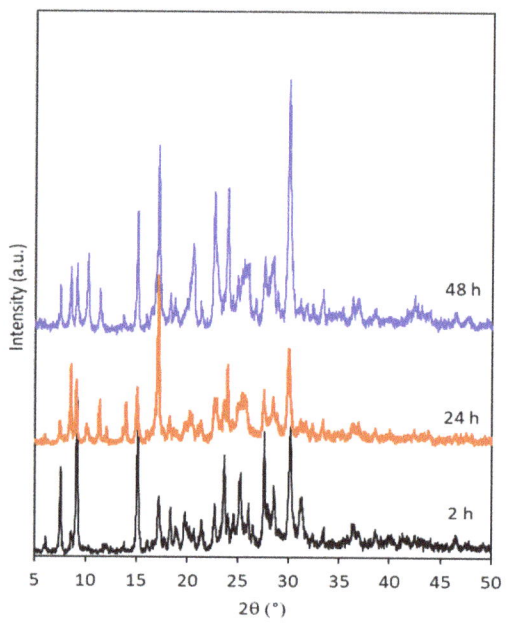

Figure 5. Phase transformations in slurry experiments in (**a**) water or (**b**) in acetonitrile.

Furthermore, the physical stability of the anhydrous oxalate salts against humidity was studied. These solids were stored in a desiccator at 75% RH using a saturated NaCl salt solution for 7 days at 40 °C and for 1 month at room temperature. After these periods the solids were again analyzed by PXRD to confirm the phase of the solids. The analysis of the solids indicated that both anhydrous oxalate salts were stable under the two stability conditions tested (Figure S9).

3.5. Computational Studies

The theoretical study is firstly focused on analyzing electronic nature of the 9ETADE ring and, more importantly, how changes upon protonation and formation of the corresponding salts with hydrogenoxalate and oxalate. Secondly, we have evaluated the H-bonding networks that are formed in the solid state of compounds **1** and **3**, as described above in Figures 3a and 4a.

Figure 6 shows the MEP surfaces of 9ETADE and its salts. In neutral 9ETADE, the MEP minimum is located at N1 (see Scheme 1a for numbering of atoms) in line with the protonation site observed in compounds **1** and **3**. The MEP values at N3 and N7 are similar (−25.1 and −24.7 kcal/mol, respectively). The MEP maximum is located at the exocyclic NH_2 group (+26.9 kcal/mol). The MEP is also large and positive (+20.0 kcal/mol) at the aromatic CH group of the five membered ring. The MEP surface of the 9-ethyladeninium–hydrogenoxalate salt extracted from the solid state of compound **1** is represented in Figure 6b. The protonation increases the MEP values at the NH_2 (+59.8 kcal/mol) and CH (+48.9 kcal/mol) groups with respect to the neutral 9ETADE, as expected. It is interesting to note that the MEP at the OH group is large and positive, despite the negative charge of hydrogenoxalate moiety. The MEP minimum is located at the O-atom of hydrogenoxalate (−81.6 kcal/mol). For the 9-ethyladeninium oxalate salt (Figure 6c), the MEP minimum is located at the oxalate anion, as expected (−77.2 kcal/mol) and the maximum at the exocyclic NH_2 group (+36.7 kcal/mol). It is worth mentioning that in this salt, the MEP value at the

CH group of the five membered ring is very large, similar to that at the NH$_2$ group, thus revealing an enhanced ability to establish H-bonds, as further discussed below.

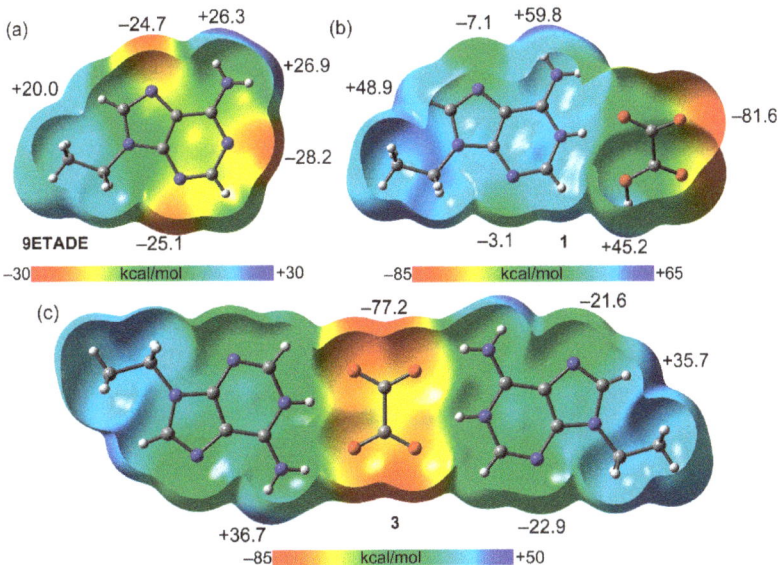

Figure 6. MEP surfaces (isovalue 0.01 a.u.) of 9ETADE (**a**), 9-ethyldeninimum hydrogenoxalate salt (**b**) and 9-ethyldeninimum oxalate salt (**c**) at the PBE0-D3/def2-TZVP level of theory. The energies at selected points of the surfaces are given in kcal/mol.

In the solid state of compound **1**, the protonated 9ETADE rings form centrosymmetric $R_2^2(10)$ H-bonded synthons, as highlighted in Figure 7, which are surrounded by four hydrogenoxalate monoanions (see Section 3.2 for distances and other geometric features). The distribution of bond and ring critical points (CPs, red and yellow spheres, respectively) and bond paths is represented in Figure 7 for the whole assembly, where the superimposed noncovalent interaction plot (NCIPlot) index isosurfaces are also represented. The NCIplot index method is conveniently used to characterize noncovalent interactions in real space. It also reveals the attractive or repulsive nature of the interactions by using a color scale. The QTAIM distribution shows that the two symmetrically equivalent N6–H7···N7 bonds that generate the $R_2^2(10)$ synthon are characterized by bond CPs and bond paths connecting the H-atoms to the N-atoms. Moreover, they are also characterized by small and blue NCIplot isosurfaces, revealing a moderately strong interaction. The formation of this assembly agrees well with the MEP analysis commented above that evidences the strong H-bond donor ability of the NH$_2$ group in the salts. The dissociation energy of each individual H-bond has been computed using the potential energy density values at the bond CPs (V_r) and the equation proposed by Espinosa et al. ($E_{dis} = -0.5 \times V_r$) [39]. The E_{dis} values are indicated in Figure 7 next to the bond CPs in red, showing that the binding energy of the $R_2^2(10)$ synthon is 7.8 kcal/mol, in line with other studies related to the energetics of Hoogsteen base pair [41,42]. Regarding the H-bonds of the $R_2^2(10)$ dimer with the counterions, the combined QTAIM/NCIplot analysis shows a quite intricate H-bonding network where each hydrogenoxalate anion establishes three H-bonds with the base pair, each one characterized by the corresponding bond CP and bond path. The strongest one corresponds to the charge assisted H-bond (N1$^+$–H···O, 7.3 kcal/mol) and the rest of H···O contacts range from 1.4 to 2.6 kcal/mol. In line with the MEP surface analysis, the CH group also participates in the H-bonding network with a moderately strong interaction energy (2.4 kcal/mol), comparable to other NH···O contacts. The total dissociation energy is

42.4 kcal/mol, thus evidencing the importance of the H-bonding network on the solid-state architecture of compound **1**.

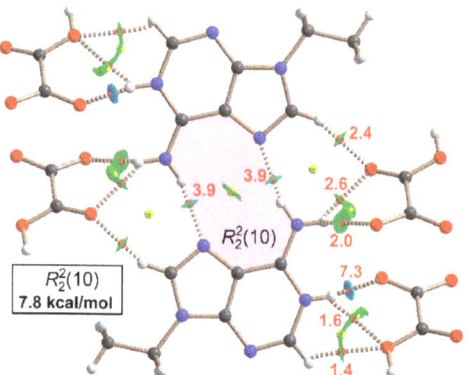

Figure 7. QTAIM distribution of bond and ring critical points (CPs, red and yellow spheres, respectively) corresponding to the H-bonds in the $R_2^2(10)$ synthons and its interaction with four surrounding $[C_2O_4H]^-$ anions. The dissociation energies of the H-bonds are indicated next to the bond CPs. Superimposed NCIplot isosurfaces [s = 0.5, cut-off = 0.04 a.u., color scale −0.04 a.u. (blue) ≤ (sign$\lambda_2)\rho$ ≤ 0.04 a.u. (red)] are also represented.

In the literature, the ability of adenine derivatives and cocrystals to form $R_2^2(10)$ homodimers or a variety of heterodimers via the Hoogsteen face has been described and energetically evaluated using the QTAIM analysis [41–43]. For instance, the reported energies associated with NH···N H-bonds in $R_2^2(10)$ ethyl-adenine homodimers in its co-crystals with malonic and fumaric acid are 3.6 and 4.0 kcal/mol [18], respectively. These values are comparable to those obtained for compound **1**. Similar interaction energies have been also reported for Hoogsteen (adenine)···Watson–Crick (N^7-pyrimidyladenine) heterodimers (3.7 and 4.2 kcal/mol) [44].

Figure 8 shows a similar QTAIM/NCIplot analysis performed for compound **3**. In this case, the $R_2^2(10)$ synthon via the Hoogsteen face is not formed. Each oxalate anion is surrounded by four 9-ethyladeninium cations establishing a network of H-bonding, characterized by the corresponding bond CPs, bond paths and blue NCIplot isosurfaces. The individual dissociated energies shown in Figure 8 reveal that two strong H-bonds are formed. One corresponds to the charge assisted HB (N1$^+$–H···O, 7.1 kcal/mol) and the other one corresponds to the N6–H6···O H-bond (6.1 kcal/mol) formed through the Watson–Crick face. The N6–H6···O H-bond formed through the Hoogsteen face is weaker (4.5 kcal/mol). These values are comparable to previously reported energies for adenine co-crystals with several carboxylic acids [18]. The QTAIM analysis also reveals the existence of a much weaker H-bond established between the aromatic CH bond of the six membered ring and N7 (2.0 kcal/mol), in line with the MEP surface plot depicted in Figure 6c. The total dissociation energy of this H-bonding network is very large (40.0 kcal/mol), specially taking into consideration that involves only one oxalate dianion, thus providing a great stability to compound **3**.

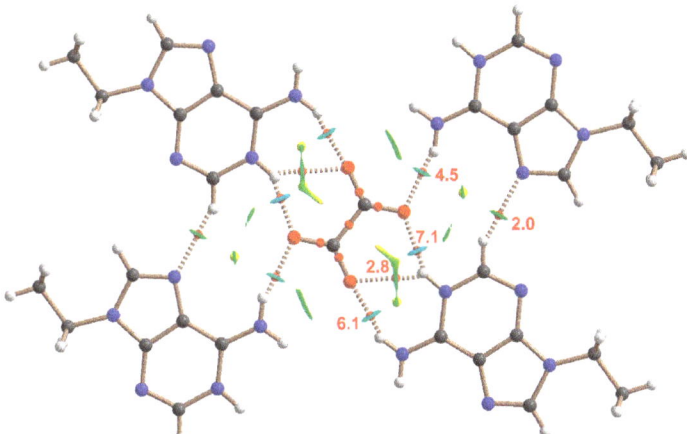

Figure 8. QTAIM distribution of bond and ring critical points (CPs, red and yellow spheres, respectively) corresponding to the pentameric assembly of compound **3**. The dissociation energies of the H-bonds are indicated next to the bond CPs. Superimposed NCIplot isosurfaces [s = 0.5, cut-off = 0.04 a.u., color scale −0.04 a.u. (blue) ≤ (signλ$_2$)ρ ≤ 0.04 a.u. (red)] are also represented.

4. Concluding Remarks

The mixture of 9-ethyladenine and oxalic acid have, surprisingly, resulted in the most productive nucleobase or derivative up to the authors' knowledge. For most of the studied nucleobases or their derivatives found in the literature (adenine, n-benzyladenine, caffeine, theobromine, theophylline or even lamivudine), the salt/cocrystal in a 2:1 ratio is the most common solid form. In this work, we have described up to six different multicomponent solids including two molar ratios and anhydrous/solvated forms using mechanochemistry, slurry or crystallization experiments. Once again, it reflects the importance of the combination of all these techniques as complementary tools besides to computational methods to fully understand the solid state landscape of multicomponent pharmaceutical solids for a determined API or model compound.

The obtained solids have been characterized and the crystal structures for the two anhydrous forms have been solved. The H-bonding networks observed in the solid-state of both compounds have been analyzed using DFT calculations combined with the QTAIM methodology. The energy of each individual H-bond has been computed, showing the relevance of the $R_2^2(10)$ centrosymmetric dimer in **1**. The dissociation energies of the charge assisted H-bonds (N1$^+$–H···O) are similar in both compounds, which are the strongest H-bonds (~7 kcal/mol) formed in the solid state. The QTAIM analysis also evidences the relevance of CH···O interactions (up to 2.4 kcal/mol) as ancillary interactions in both compounds.

Supplementary Materials: The following are available online at https://www.mdpi.com/article/10.3390/cryst12010089/s1, Experimental methods for solution syntheses and grinding screening. Figure S1: TGA-DSC traces of 9-ethyladenine-oxalic acid (1:1) salt (1). Figures S2: TGA-DSC traces of 9-ethyladenine-oxalic acid (2:1) salt (3). Figure S3: TGA-DSC traces of 9-ethyladenine-oxalic acid hydrated (1:1) salt (2). Figure S4: TGA-DSC traces of 9-ethyladenine-oxalic acid hydrated salt (4). Figures S5: TGA-DSC traces of 9-ethyladenine-oxalic acid hydrated (1:1) salt (5). Figure S6: TGA-DSC traces of 9-ethyladenine-oxalic acid hydrated (1:1) hydrated salt (6). Figure S7: Ortep images of compounds (a) 1 and (b) 3. Table S1: Hydrogen bond details for 9-ethyladenine-oxalic acid (1:1) salt (1). Table S2: Hydrogen bond details for 9-ethyladenine-oxalic acid (2:1) salt (3). Figure S8: Diffractograms of initial and undissolved solids for compounds 9ETADE-OXA (2:1) (3), 9ETADE-

OXA (1:1) (1) and 9ETADE-SUC (1:1) salt. Figure S9: PXRD patterns of the anhydrous 9ETADE-OXA salts under different temperature and relative humidity conditions.

Author Contributions: M.B.-O. prepared 9-ethyladenine. M.B. designed the experiments, characterized the samples and obtained the single crystals. M.B.-O. performed the data collection and the resolution of the crystal structures. A.F. performed the D.F.T. study. M.B., A.F. and E.M. designed the concept. E.M. supervised the work. All authors wrote and revised the manuscript. All authors have read and agreed to the published version of the manuscript.

Funding: M.B. and E.M. are grateful to the Severo Ochoa FunFuture project (MICINN, CEX2019-917S) and Generalitat de Catalunya (2017SGR1687). M.B.-O. thanks the Vice-Rector for Research and International Relations of the University of the Balearic Islands for the financial support in setting up the single-crystal X-ray diffraction facility. A.F. thanks the MICIU/AEI from Spain for financial support (Projects CTQ2017-85821-R and PID2020-115637GB-I00, Feder funds).

Institutional Review Board Statement: Not applicable.

Informed Consent Statement: Not applicable.

Data Availability Statement: Not applicable.

Acknowledgments: We thank R. Frontera from the Centre de Tecnologies de la Informació (CTI) at the UIB for computational facilities. The authors also thank the powder diffraction and thermal analysis services from the ICMAB.

Conflicts of Interest: The authors declare no conflict of interest.

References

1. Stahl, P.H.; Wermuth, C.G. (Eds.) *Handbook of Pharmaceutical Salts: Properties, Selection and Use*; Wiley-VCH: Zurich, Switzerland, 2008.
2. Harrison, W.T.A.; Yathirajan, H.S.; Bindya, S.; Anilkumar, H.G. Devaraju Escitalopram oxalate: Co-existence of oxalate dianions and oxalic acid molecules in the same crystal. *Acta Crystallogr. Sect. C Cryst. Struct. Commun.* **2007**, *63*, o129–o131. [CrossRef]
3. Åaslund, B.L.; Aurell, C.J.; Bohlin, M.H.; Sebhatu, T.; Ymen, B.I.; Healy, E.T.; Jensen, D.R.; Jonaitis, D.T.; Parent, S. Crystalline Naloxol-Peg Conjugate. Patent WO2012044293A1, 5 April 2012.
4. Gelbrich, T.; Langes, C.; Stefinovic, M.; Griesser, U.J. Naloxegol hydrogen oxalate displaying a hydrogen-bonded layer structure. *Acta Crystallogr. Sect. E Crystallogr. Commun.* **2018**, *74*, 474–477. [CrossRef] [PubMed]
5. Paulekuhn, G.S.; Dressman, J.B.; Saal, C. Trends in Active Pharmaceutical Ingredient Salt Selection based on Analysis of the Orange Book Database. *J. Med. Chem.* **2007**, *50*, 6665–6672. [CrossRef] [PubMed]
6. Owoyemi, B.C.D.; da Silva, C.C.P.; Diniz, L.F.; Souza, M.S.; Ellena, J.; Carneiro, R.L. Fluconazolium oxalate: Synthesis and structural characterization of a highly soluble crystalline form. *CrystEngComm* **2019**, *21*, 1114–1121. [CrossRef]
7. Chen, Y.; Li, L.; Yao, J.; Ma, Y.-Y.; Chen, J.-M.; Lu, T.-B. Improving the Solubility and Bioavailability of Apixaban via Apixaban–Oxalic Acid Cocrystal. *Cryst. Growth Des.* **2016**, *16*, 2923–2930. [CrossRef]
8. Dichiarante, E.; Curzi, M.; Giaffreda, S.L.; Grepioni, F.; Maini, L.; Braga, D. Crystal forms of the hydrogen oxalate salt of o-desmethylvenlafaxine. *J. Pharm. Pharmacol.* **2015**, *67*, 823–829. [CrossRef]
9. Perumalla, S.R.; Sun, C.C. Design and Preparation of a 4:1 Lamivudine–Oxalic Acid CAB Cocrystal for Improving the Lamivudine Purification Process. *Cryst. Growth Des.* **2014**, *14*, 3990–3995. [CrossRef]
10. Available online: https://www.fda.gov/news-events/press-announcements/fda-hold-advisory-committee-meeting-discuss-merck-and-ridgebacks-eua-application-covid-19-oral (accessed on 1 December 2021).
11. Sridhar, B.; Ravikumar, K.; Varghese, B. Supramolecular hydrogen-bonded networks in adeninediiium hemioxalate chloride and adeninium semioxalate hemi(oxalic acid) monohydrate. *Acta Crystallogr. Sect. C Cryst. Struct. Commun.* **2009**, *65*, o202–o206. [CrossRef]
12. McHugh, C.; Erxleben, A. Supramolecular Structures and Tautomerism of Carboxylate Salts of Adenine and Pharmaceutically Relevant N6-Substituted Adenine. *Cryst. Growth Des.* **2011**, *11*, 5096–5104. [CrossRef]
13. Trask, A.V.; Motherwell, W.D.S.; Jones, W. Pharmaceutical Cocrystallization: Engineering a Remedy for Caffeine Hydration. *Cryst. Growth Des.* **2005**, *5*, 1013–1021. [CrossRef]
14. Otsuka, Y.; Ito, A.; Takeuchi, M.; Tanaka, H. Dry Mechanochemical Synthesis of Caffeine/Oxalic Acid Cocrystals and Their Evaluation by Powder X-ray Diffraction and Chemometrics. *J. Pharm. Sci.* **2017**, *106*, 3458–3464. [CrossRef] [PubMed]
15. Fischer, F.; Scholz, G.; Batzdorf, L.; Wilke, M.; Emmerling, F. Synthesis, structure determination, and formation of a theobromine: Oxalic acid 2:1 cocrystal. *CrystEngComm* **2014**, *17*, 824–829. [CrossRef]
16. Zhang, S.; Rasmuson, C. The theophylline–oxalic acid co-crystal system: Solid phases, thermodynamics and crystallisation. *CrystEngComm* **2012**, *14*, 4644–4655. [CrossRef]
17. Trask, A.V.; Motherwell, W.D.S.; Jones, W. Physical stability enhancement of theophylline via cocrystallization. *Int. J. Pharm.* **2006**, *320*, 114–123. [CrossRef]

18. Roselló, Y.; Benito, M.; Bagués, N.; Martínez, N.; Moradell, A.; Mata, I.; Galcerà, J.; Barceló-Oliver, M.; Frontera, A.; Molins, E. 9-Ethyladenine: Mechanochemical Synthesis, Characterization, and DFT Calculations of Novel Cocrystals and Salts. *Cryst. Growth Des.* **2020**, *20*, 2985–2997. [CrossRef]
19. García-Raso, A.; Fiol, J.J.; Bádenas, F.; Solans, X.; Font-Bardia, M. Reaction of trimethylene–bisadenine with d10 divalent cations. *Polyhedron* **1999**, *18*, 765–772. [CrossRef]
20. *SADABS, Bruker-AXS*; Version 1; Bruker AXS Inc.: Madison, WI, USA, 2004.
21. Farrugia, L.J. WinGX suite for small-molecule single-crystal crystallography. *J. Appl. Crystallogr.* **1999**, *32*, 837–838. [CrossRef]
22. Sheldrick, G.M. Crystal structure refinement with SHELXL. *Acta Crystallogr.* **2015**, *C71*, 3–8. [CrossRef]
23. Sheldrick, G.M. *SHELXL-2017/1, Program for the Solution of Crystal Structures*; University of Göttingen: Göttingen, Germany, 2017.
24. Spek, A.L. Structure validation in chemical crystallography. *Acta Crystallogr.* **2009**, *D65*, 148–155. [CrossRef]
25. Ahuja, D.; Svärd, M.; Rasmuson, C. Investigation of solid–liquid phase diagrams of the sulfamethazine–salicylic acid co-crystal. *CrystEngComm* **2019**, *21*, 2863–2874. [CrossRef]
26. Alvarez-Lorenzo, C.; Castiñeiras, A.; Frontera, A.; García-Santos, I.; González-Pérez, J.M.; Niclós-Gutiérrez, J.; Rodríguez-González, I.; Vílchez-Rodríguez, E.; Zaręba, J.K. Recurrent motifs in pharmaceutical cocrystals involving glycolic acid: X-ray characterization, Hirshfeld surface analysis and DFT calculations. *CrystEngComm* **2020**, *22*, 6674–6689. [CrossRef]
27. Rockland, L.B. Saturated Salt Solutions for Static Control of Relative Humidity between 5° and 40° C. *Anal. Chem.* **1960**, *32*, 1375–1376. [CrossRef]
28. Frisch, M.J.; Trucks, G.W.; Schlegel, H.B.; Scuseria, G.E.; Robb, M.A.; Cheeseman, J.R.; Scalmani, G.; Barone, V.; Mennucci, B.; Petersson, G.A.; et al. *Gaussian16*; Revision A.03; Gaussian Inc.: Wallingford, CT, USA, 2016.
29. Adamo, C.; Barone, V. Toward reliable density functional methods without adjustable parameters: The PBE0 model. *J. Chem. Phys.* **1999**, *110*, 6158–6170. [CrossRef]
30. Grimme, S.; Antony, J.; Ehrlich, S.; Krieg, H. A consistent and accurate ab initio parametrization of density functional dispersion correction (DFT-D) for the 94 elements H-Pu. *J. Chem. Phys.* **2010**, *132*, 154104. [CrossRef] [PubMed]
31. Weigend, F. Accurate Coulomb-fitting basis sets for H to Rn. *Phys. Chem. Chem. Phys.* **2006**, *8*, 1057–1065. [CrossRef] [PubMed]
32. Gomila, R.M.; Frontera, A. Metalloid Chalcogen–pnictogen σ-hole bonding competition in stibanyl telluranes. *J. Organomet. Chem.* **2021**, *954–955*, 122092. [CrossRef]
33. Mertsalov, D.F.; Gomila, R.M.; Zaytsev, V.P.; Grigoriev, M.S.; Nikitina, E.V.; Zubkov, F.I.; Frontera, A. On the Importance of Halogen Bonding Interactions in Two X-ray Structures Containing All Four (F, Cl, Br, I) Halogen Atoms. *Crystals* **2021**, *11*, 1406. [CrossRef]
34. García-Rubiño, M.; Matilla-Hernández, A.; Frontera, A.; Lezama, L.; Niclós-Gutiérrez, J.; Choquesillo-Lazarte, D. Dicopper(II)-EDTA Chelate as a Bicephalic Receptor Model for a Synthetic Adenine Nucleoside. *Pharmaceuticals* **2021**, *14*, 426. [CrossRef]
35. Le, A.T.; Tran, V.T.T.; Le, D.T.; Gomila, R.M.; Frontera, A.; Zubkov, F.I. Synthesis, X-ray characterization and theoretical study of all-cis 1,4:2,3:5,8:6,7-tetraepoxynaphthalenes: On the importance of the through-space α-effect. *CrystEngComm* **2021**, *23*, 7462–7470. [CrossRef]
36. Bader, R.F.W. A Bond Path: A Universal Indicator of Bonded Interactions. *J. Phys. Chem. A* **1998**, *102*, 7314–7323. [CrossRef]
37. Contreras-Garcia, J.; Johnson, E.R.; Keinan, S.; Chaudret, R.; Piquemal, J.-P.; Beratan, D.; Yang, W. NCIPLOT: A Program for Plotting Noncovalent Interaction Regions. *J. Chem. Theory Comput.* **2011**, *7*, 625–632. [CrossRef] [PubMed]
38. Keith, T.A. *AIMALL*; Version 19.10.12; TK Gristmill Software: Overland Park, KS, USA, 2019.
39. Espinosa, E.; Molins, E.; Lecomte, C. Hydrogen bond strengths revealed by topological analyses of experimentally observed electron densities. *Chem. Phys. Lett.* **1998**, *285*, 170–173. [CrossRef]
40. Carvalho, P.S.; Diniz, L.F.; Tenorio, J.C.; Souza, M.S.; Franco, C.H.J.; Rial, R.; De Oliveira, K.R.W.; Nazario, C.E.D.; Ellena, J. Pharmaceutical paroxetine-based organic salts of carboxylic acids with optimized properties: The identification and characterization of potential novel API solid forms. *CrystEngComm* **2019**, *21*, 3668–3678. [CrossRef]
41. García-Raso, A.; Terrón, A.; López-Zafra, A.; García-Viada, A.; Barta, A.; Frontera, A.; Lorenzo, J.; Rodríguez-Calado, S.; Vázquez-López, E.M.; Fiol, J.J. Crystal structures of N6-modified-amino acid related nucleobase analogs (II): Hybrid adenine-β-alanine and adenine-GABA molecules. *New J. Chem.* **2019**, *43*, 9680–9688. [CrossRef]
42. García-Raso, A.; Terrón, A.; Bauzá, A.; Frontera, A.; Molina, J.J.; Vázquez-López, E.M.; Fiol, J.J. Crystal structures of N6-modified-aminoacid/peptide nucleobase analogs: Hybrid adenine-glycine and adenine-glycylglycine molecules. *New J. Chem.* **2018**, *42*, 14742–14750. [CrossRef]
43. García-Raso, A.; Terrón, A.; Balle, B.; López-Zafra, A.; Frontera, A.; Barceló-Oliver, M.; Fiol, J.J. Crystal structures of N6-modified-amino acid nucleobase analogs(iii): Adenine-valeric acid, adenine-hexanoic acid and adenine-gabapentine. *New J. Chem.* **2020**, *44*, 12236–12246. [CrossRef]
44. Martínez, D.; Pérez, A.; Cañellas, S.; Silió, I.; Lancho, A.; García-Raso, A.; Fiol, J.J.; Terrón, A.; Barceló-Oliver, M.; Orte-ga-Castro, J.; et al. Synthesis, reactivity, X-ray characterization and docking studies of N7/N9-(2-pyrimidyl)-adenine derivatives. *J. Inorg. Biochem.* **2020**, *203*, 110879. [CrossRef]

Article

Multicomponent Solids of DL-2-Hydroxy-2-phenylacetic Acid and Pyridinecarboxamides

Alfonso Castiñeiras [1,*], Antonio Frontera [2], Isabel García-Santos [1], Josefa M. González-Pérez [3], Juan Niclós-Gutiérrez [3] and Rocío Torres-Iglesias [1]

[1] Department of Inorganic Chemistry, Faculty of Pharmacy, University of Santiago de Compostela, 15782 Santiago de Compostela, Spain; isabel.garcia@usc.es (I.G.-S.); rocio.torres.iglesias@rai.usc.es (R.T.-I.)
[2] Department of Química, Universitat de les Illes Balears, Crta. De Valldemossa km 7.5, 07122 Palma de Mallorca, Spain; toni.frontera@uib.es
[3] Department of Inorganic Chemistry, Faculty of Pharmacy, University of Granada, 18071 Granada, Spain; jmgp@ugr.es (J.M.G.-P.); jniclos@ugr.es (J.N.-G.)
* Correspondence: alfonso.castineiras@usc.es

Abstract: We prepared cocrystals of DL-2-Hydroxy-2-phenylacetic acid (**D, L-H$_2$ma**) with the pyridinecarboxamide isomers, picolinamide (**pic**) and isonicotinamide (**inam**). They were characterized by elemental analysis, single crystal and powder X-ray, IR spectroscopy and ^{1}H and ^{13}C NMR. The crystal and molecular structures of (**pic**)-(**D-H$_2$ma**) (**1**), (**nam**)-(**L-H$_2$ma**) (**2**) and (**inam**)-(**L-H$_2$ma**) (**3**) were studied. The crystal packing is stabilized primarily by hydrogen bonding and in some cases through π-π stacking interactions. The analysis of crystal structures reveals the existence of the characteristic heterosynthons with the binding motif $R_2^2(8)$ (primary amide–carboxylic acid) between pyridinecarboxamide molecules and the acid. Other synthons involve hydrogen bonds such as O-H$_{(carboxyl)}$···N$_{(pyridine)}$ and O-H$_{(hydroxyl)}$···N$_{(pyridine)}$ depending on the isomer. The packing of 1 and 3 is formed by tetramers, for whose formation a crystallization mechanism based on two stages is proposed, involving an amide–acid (**1**) or amide–amide (**3**) molecular recognition in the first stage and the formation of others, and interdimeric hydrogen bonding interactions in the second. The thermal stability of the cocrystals was studied by differential scanning calorimetry and thermogravimetry. Further studies were conducted to evaluate other physicochemical properties of the cocrystals in comparison to the pure coformers. Density-functional theory (DFT) calculations (including NCIplot and QTAIM analyses) were performed to further characterize and rationalize the noncovalent interactions.

Keywords: pyridinecarboxamides; cocrystals; mandelic acid; X-ray structure; DFT calculations

Citation: Castiñeiras, A.; Frontera, A.; García-Santos, I.; González-Pérez, J.M.; Niclós-Gutiérrez, J.; Torres-Iglesias, R. Multicomponent Solids of DL-2-Hydroxy-2-phenylacetic Acid and Pyridinecarboxamides. *Crystals* 2022, 12, 142. https://doi.org/10.3390/cryst12020142

Academic Editor: Klaus Merz

Received: 10 December 2021
Accepted: 5 January 2022
Published: 20 January 2022

Publisher's Note: MDPI stays neutral with regard to jurisdictional claims in published maps and institutional affiliations.

Copyright: © 2022 by the authors. Licensee MDPI, Basel, Switzerland. This article is an open access article distributed under the terms and conditions of the Creative Commons Attribution (CC BY) license (https://creativecommons.org/licenses/by/4.0/).

1. Introduction

Crystal engineering is the rational design of functional molecular solids from neutral or ionic building blocks, using intermolecular interactions in the design strategy [1]. This field has its origins in organic chemistry and in physical chemistry. The expansion of crystal engineering during the last years as a research field parallels significant interest in the origin and nature of intermolecular interactions and their use in the design and preparation of new crystalline structures [2].

The concept of crystal engineering, mainly cocrystal, is gaining an extensive interest of pharmaceutical researchers of both academia and industry during the last decade [3], the prominent reason being its ability to enhance the physicochemical and biopharmaceutical properties of active pharmaceutical ingredients without altering chemical structure, thus maintaining its therapeutic activity. With the new guidelines issued by United States Food and Drug Administration and European Medicines Agency for the regulatory aspect of cocrystal, the development of pharmaceutical cocrystal has gained a high impetus [4].

A supramolecular synthesis is used to prepare cocrystals, and the design of homo and supramolecular heterosynthons is particularly one of the most exploited [5,6]. Although the preparation of cocrystals does not involve great complexity, the selection of the solvent can be critical in obtaining a particular crystal phase of cocrystal. The role of the solvent in the nucleation of crystals and cocrystals is still far from being completely understood.

The three isomers of pyridinecarboxamide: 2-pyridine carboxamide or picolinamide (**pic**), 3-pyridinecarboxamide or nicotinamide (**nam**) and 4-pyridinecarboxamide or isonicotinamide (**inam**) (Scheme 1), are a class of medicinal agents that can be classified as GRAS compounds (generally regarded as safe). Nicotinamide and isonicotinamide are popular cocrystal formers, **nam** is vitamin B3 and therefore of pharmaceutical relevance [7], whilst isonicotinamide is one of the most effectively used cocrystallizing compounds [8], as the pyridine N atom of the isonicotinamide molecule readily acts as a hydrogen bond acceptor when faced with good hydrogen bond donors such as carboxylic acids and alcohols [9]. In fact, the carboxylic acid···pyridine hydrogen bond has been identified as a robust yet versatile hydrogen bond and persists even in the presence of other good donors [10]. Cocrystals of picolinamide are rarely seen in the literature, despite being a structural isomer of **nam** and **inam** and a strong inhibitor of poly(ADP-ribose)synthetase [11], showing important biological activity with a coenzyme called NAD (nicotinamide adenine dinucleotide), which plays important roles in more than 200 amino acid and carbohydrate metabolic reactions [12]. Apart from pharmaceutical value, in general, pyridinecarboxamides are excellent cocrystallizing compounds. The amide group features two hydrogen bond donors and two lone pairs on the carbonyl O atom. A second hydrogen bond acceptor is the lone pair on the N atom of the pyridine ring. Consequently, these molecules are very versatile for a variety of hydrogen bonding interactions, especially in pharmaceutical cocrystals [13]. The crystal and molecular structures of the three isomers have been the subject of recent studies, and from the crystal structure point of view, all isomer compounds exhibit polymorphism. Two polymorphs are known of picolinamide [14] nicotinamide, which is a highly polymorphic compound with nine solved single-crystal structures [15], and isonicotinamide has six polymorphs in monoclinic and orthorhombic forms [16–19].

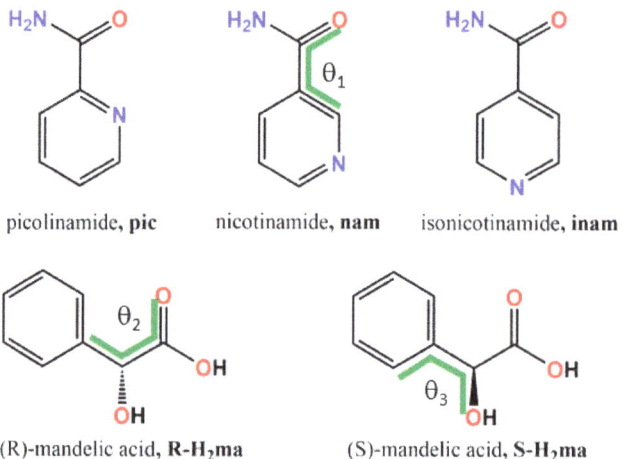

Scheme 1. Molecular diagrams of pyridinecarboxamides and mandelic acid. θ_1, θ_2 and θ_3 indicate angles of torsion.

DL-Mandelic acid (Scheme 1) is a useful precursor to various drugs, for example homatropine and cyclandelate, which are esters of mandelic acid, and it is also known to have antibacterial properties [20]. Generally, the profile of **DL-H$_2$ma** allows us to envisage this compound as an excellent coformer for cocrystals with the aforementioned carboxamides. Indeed, given that **DL-H₂ma** is a substituted carboxylic acid containing a hydroxyl group on the adjacent carbon, it also possesses a set of sites capable of hydrogen bonding, both of donor and acceptor character.

For racemate of mandelic acid **DL-H₂ma** and its enantiomers **L-H₂ma** and **D-H₂ma**, different polymorphs are known. Racemic **DL-H₂ma** occurs as orthorhombic form I and also as polymorph II, monoclinic, metastable at normal conditions [21,22]. From **D-H₂ma** only one monoclinic form is known [23] and for **L-H₂ma** two polymorphs are known, both monoclinic [22].

Taking into account the previous considerations, the main objective of this work is the preparation, characterization, study of physicochemical properties, and identification of recurrent supramolecular patterns, within a new set of multicomponent pharmaceutical crystals that involve the three isomers of pyridinecarboxamide with DL-mandelic acid as coformer (Scheme 1), as well as to evaluate the effect of position isomerism of cocrystal formers on the formation and robustness of the supramolecular structures and subsequent physicochemical properties of cocrystals.

2. Experimental Part

2.1. Materials and Physical Measurements

DL-Mandelic acid and the pyridinecarboxamide isomers were purchased from Sigma-Aldrich (Sigma-Aldrich. Inc., Tres Cantos, Madrid, Spain). Commercially available solvents were used as received without further purification. Melting points were determined on a Büchi melting point apparatus (Büchi Labortechnik AG-Flawil, Switzerland) and are uncorrected. Elemental analyses for carbon, hydrogen and nitrogen were performed with a Fisons-Carlo Erba 1108 microanalyser (CARLO ERBA Reagents SAS, Chaussée du Vexin, France). ^{1}H NMR and ^{13}C NMR spectra in DMSO-d_6 were run on a Varian Mercury 300 instrument (Varian Medical Systems, Inc., Palo Alto, CA, USA), using TMS as the internal reference. IR spectra were recorded as KBr pellets (4000–400 cm^{-1}) on a Bruker IFS-66v spectrophotometer (Bruker Corporation, Billerica, MA, USA). TGA experiments were carried out on a Shimadzu Thermobalance TGA-DTG-50H Instrument (TA Instruments, New Castle, DE, EE. UU.) from room temperature to 700 °C in a flow of air (100 mL min^{-1}) and series of FTIR spectra (20–30 per sample) of evolved gases were recorded using a coupled FT–IR Nicolet Magma 550 spectrophotometer (Thermo Fisher Scientific, Waltham, MA, USA). Differential scanning calorimetry (DSC) was conducted on a DSC Q100 apparatus of TA Instruments. Accurately weighed samples (1–2 mg) were placed in hermetically sealed aluminum crucibles (40 mL) and scanned from 0 to 350 °C at a heating rate of 10 °C min^{-1} under a dry nitrogen atmosphere. Powder X-ray diffraction (XRPD) patterns were collected on a Philips PW1710 (Philips Engineering Solutions, Aachen, Germany) with a Cu-Kα radiation (1.5406 Å). The tube voltage and amperage were set at 40 kV and 30 mA, respectively. Each sample was scanned between 2 and 50° 2θ with a step size of 0.02°. The instrument was previously calibrated using a silicon standard.

2.2. Cocrystal Screening

Compounds were prepared by cocrystallization via solvent-drop grinding: Stoichiometric amounts of **DL-H₂ma** with **pic**, or **inam** were ground with a mortar and pestle for ca. 5–7 min with the addition of 10 μL of Ethyl acetate per 50 mg of cocrystal formers. The clear, nonsaturated resulting solutions were poured into a 5 mL vial and left to evaporate slowly under ambient conditions until cocrystals formed. The single crystals of (**pic**)-(**D-H₂ma**) (**1**) and (**inam**)-(**L-H₂ma**) (**3**), suitable for X-ray diffraction studies, were obtained in 2–15 days from Ethyl acetate.

Preparation of the cocrystals was also attempted with other solvents, such as dichloromethane, formamide, DMF, chloroform, acetonitrile, isopropyl alcohol, cyclohexane, ethanol, methanol, CCl₄, THF and toluene, but the yield was lower.

2.3. Cocrystal Synthesis

(pic)-(D-H$_2$ma) (**1**). DL-mandelic acid (0.152 g, 1.0 mmol) and picolinamide (0.122 g, 1.0 mmol). Ethyl acetate (3 mL). Colorless crystals after one day. M.p. (°C): 88–92. Elemental analysis: Found: C, 59.72; H, 5.10; N, 10.27. Calculated (%) for C$_{14}$H$_{14}$N$_2$O$_4$: C, 61.31; H, 5.45; N, 10.21. IR (ν_{max}/cm^{-1}): 3471–3435 ν(NH$_2$), ν(OH), 3282 ν(NH$_2$), 1695–1633 ν(C=O), 1589, 1570, 1418 ν(CN), 1293, 1269, 1187, 1107, 1066–1012 ν(C-O), 995, 755, 725, 697, 645–616 α(CCC), 600. ^{1}H NMR (DMSO-d$_6$, ppm): 12.50 (br, 1H, COOH), 8.62 (m, 1H, py), 8.12 (s, 1H, NH$_2$), 8.05–7.97(m, 2H, py), 7.64 (s, 1H, NH$_2$), 7.65–7.58 (m, 5H, ring), 5.86 (s, 1H, CH), 5.02 (s, 1H, OH). ^{13}C NMR (DMSO-d$_6$, ppm): 174.3 (C1), 166.3 (C11), 150.4 (C12), 140.4 (C3), 137.8 (C14), 128.3 (C3, C7), 127.8 (C6), 126.8 (C4, C8), 126.6 (C15), 122.0 (C13), 72.4 (C2).

(inam)-(L-H$_2$ma) (**3**). DL-mandelic acid (0.152 g, 1.0 mmol) and isonicotinamide (0.122 g, 1.0 mmol). Ethyl acetate (11 mL). Colorless crystals after one day. M.p. (°C): 107–114. Elemental analysis: Found: C, 59.89; H, 4.85; N, 10.55. Calculated (%) for C$_{14}$H$_{16}$N$_2$O$_4$: C, 61.31; H, 5.45; N, 10.21. IR (ν_{max}/cm^{-1}): 3424–3379 ν(NH$_2$), ν(OH), 3162 ν(NH$_2$), 1729, 1694–1604 ν(C=O), 1555, 1420–1407 ν(CN), 1303, 1227, 1188, 1065–1018 ν(C-O), 756, 737, 695, 647 α(CCC), 606. ^{1}H NMR (DMSO-d$_6$, ppm): 12.53 (br, 1H, COOH), 8.70–8.60(m, 2H, py), 8.22 (s, 1H, NH$_2$), 7.75–7.70 (m, 3H, py) 7.40 (s, 1H, NH$_2$), 7.37–7.23 (m, 5H, ring), 5.82 (s, 1H, CH), 4.99 (s, 1H, OH). ^{13}C NMR (DMSO-d$_6$, ppm): 174.3 (C1), 166.5 (C11), 150.4 (C14, C16), 140.3 (C12), 128.2 (C3, C7), 127.7 (C6), 126.7 (C4, C8), 121.5 (C13, C17), 72.4 (C2).

2.4. Solubility Determination

The water solubility of each cocrystal was determined in triplicate, and compared with that of the corresponding pyridinecarboxamides coformer. For this, a quantitative method based on the saturation of solutions at room temperature was used as follows: An excess of coformer or cocrystal was dispersed in 0.5 or 1 mL of Milli-Q water in Eppendorf tubes and left stirring for a week at room temperature. The samples were centrifuged at 12,000 rpm for 45 min, in each case the supernatant liquid was removed, and the corresponding solid was dried and weighed. This led us first to determine the solubility for **pic** (S = 177 mg/mL), **nam** (S = 500 mg/mL), **inam** (S = 192 mg/mL) and **DL-H$_2$ma** (S = 159 mg/mL), and then that of their respective cocrystals.

2.5. Theoretical Methods

Calculations of the noncovalent interactions were performed using the Gaussian-16 [24] program and the PBE0-D3/def2-TZVP level of theory [25–27]. To evaluate the interactions in the solid state, the crystallographic coordinates were used apart from the positions of the H-atoms, which were optimized at the same level. The interaction energies were calculated with correction for the basis set superposition error (BSSE) using the Boys–Bernardi counterpoise technique [28]. Bader's quantum theory "atoms in molecules" theory (QTAIM) and noncovalent interaction plot (NCIPlot) [29] were used to study the interactions discussed herein by means of the AIMall calculation package [30]. The molecular electrostatic potential surfaces were computed using the Gaussian-16 software and the 0.001 a.u. cutoff for the isosurface.

Table 1. Crystal data and structure refinement for cocrystals.

Compound	(pic)-(D-H$_2$ma) (1)	HOGGOB [31]	JILZOU01 (2a) [32]	JILZOU (2b) [33]	(inam)-(L-H$_2$ma) (3)
Empirical formula	C$_{14}$H$_{14}$N$_2$O$_4$	C$_{14}$H$_{14}$N$_2$O$_4$	C$_{14}$H$_{14}$N$_2$O$_4$	C$_{14}$H$_{14}$N$_2$O$_4$	C$_{14}$H$_{14}$N$_2$O$_4$
Formula weight	274.27	274.27	274.27	274.27	274.27
Temperature/K	100 (2)	100 (2)	100 (2)	150 (2)	100 (2)
Wavelength/Å	0.71073	1.54178	1.54178	0.71073	0.71073
Crystal system	Monoclinic	Monoclinic	Monoclinic	Monoclinic	Monoclinic
Space group	$P2_1/n$	$P2_1$	$P2_1$	$C2$	$P2_1/c$
a/Å	5.4240 (3)	5.390 (2)	5.2406 (3)	32.6557 (9)	5.2201 (8)
b/Å	26.1177 (14)	9.897 (3)	10.0477 (6)	5.4751 (10)	27.662 (4)
c/Å	9.3622 (5)	24.214 (6)	12.6006 (7)	14.9264 (5)	9.1862 (15)
α/°					
β/°	104.715 (2)	90.699 (13)	95.678 (4)	99.400 (1)	99.935 (10)
γ/°					
Volume/Å^{-3}	1282.77 (12)	1291.6 (7)	660.24 (7)	2632.9 (5)	1306.6 (4)
Z	4	4	2	8	4
Calc. density/Mg/m^3	1.420	1.410	1.380	1.384	1.394
Absorp. coefc./mm^{-1}	0.106	0.876	0.857	0.103	0.104
F(000)	576	576	288	1152	576
Crystal size	0.34 × 0.18 × 0.12	0.25 × 0.10 × 0.03	0.28 × 0.18 × 0.16	0.46 × 0.07 × 0.07	0.54 × 0.26 × 0.16
θ range/°	2.381–27.484	1.82–66.57	3.52–67.60	3.77–27.43	2.368–29.571
Limiting indices/h,k,l	−7/6, −33/33, −12/12	−6/6, −11/11, −28/28	−6/6, −9/11, −15/14	−42/41, −6/7, −15/19	−7/7, −38/38, −12/12
Absorp. correct.	Multiscan	Multiscan	Multiscan	Multiscan	Multiscan
Max. /min. transm.	1.000–0.946	1.000–0.832	1.000–0.909	1.000–0.569	1.000–0.913
Data/parameters	2926/196	4517/365	1235/183	5773/385	3655/196
Goodness-of-fit on F^2	1.026	1.066	1.011	1.122	1.097
Final R indices	$R_1 = 0.0378$, $wR_2 = 0.0871$	$R_1 = 0.0271$, $wR_2 = 0.0695$	$R_1 = 0.0430$, $wR_2 = 0.1138$	$R_1 = 0.0550$, $wR_2 = 0.1278$	$R_1 = 0.0468$, $wR_2 = 0.1149$
R indices (all data)	$R_1 = 0.0492$, $wR_2 = 0.0936$	$R_1 = 0.0287$, $wR_2 = 0.0708$	$R_1 = 0.0472$, $wR_2 = 0.1174$	$R_1 = 0.0703$, $wR_2 = 0.1374$	$R_1 = 0.0633$, $wR_2 = 0.1234$
Largest dif. peak/hole	0.305/−0.248	0.159/−0.189	0.306/−0.248	0.291/−0.298	0.444/−0.276
CCDC number	2072590	977791	904263	626647	2072591

3. Results and Discussion

The cocrystallization processes were carried out considering the pK_a of the mandelic acid and the pyridinecarboxamide isomers as coformers, having the pK_a values of 3.85 (**DL-H$_2$ma**), based on the carboxylic group [34], 2.10 (**pic**), 3.35 (**nam**) and 3.61 (**inam**), based on pyridine nitrogen [35]. These compounds were chosen to evaluate the degree of acid proton transfer to the coformers, according to the ΔpK_a rule, which can contribute to the study of the salt/cocrystal continuum and provide information related to the ability to predict and control synthesis of cocrystals that contain mandelic acid [36]. According to this rule, it is generally accepted that a salt is formed when the value of ΔpK_a is greater than 3, while a value of ΔpK_a less than 0 should lead to the formation of cocrystals [37]. The values of ΔpK_a (pK_a(protonated base)-pK_a(acid)) calculated for **pic**, **nam** and **inam** are −1.75, −0.50 and −0.24, respectively, so the formation of cocrystals should be expected.

The binary solid forms were characterized using NMR (Supplementary Information, Figures S1–S11) and IR (Supplementary Information, Figures S13 and S14) spectroscopy, powder X-ray diffraction (Supplementary Information, Figure S12), and thermal DSC technique. The (**pic**)-(**D-H$_2$ma**) (**1**) and (**inam**)-(**L-H$_2$ma**) (**3**) crystal structures were established using the single crystal X-ray diffraction technique. The crystal data, experimental details and refinement results are summarized in Table 1. Data of (**nam**)-(L-H$_2$ma) (**2**) and its crystal structure was taken from the bibliography (JILZOU01 [32]).

3.1. Crystal Structure Analysis

Cocrystallization of **DL-H$_2$ma** and **pic** in 1:1 molar ratio from ethyl acetate produced plate-shaped colorless crystals that belonged to a 1:1 cocrystal, a new polymorph that differs from the one known with a 2:2 **pic-D-H$_2$ma** ratio (HOGGOB, [31]). The crystal structure was solved in monoclinic space group $P2_1/n$. The crystallographic asymmetric unit consists of one molecule each of **D-H₂ma** and **pic** (Figure 1a). The crystal structure features an acid–amide heterosynthon $R_2^2(8)$ between **D-H₂ma** and **pic** involving O–H···O (2.547(1) Å, 168.3(2)°) and N–H···O (2.929(2) Å, 166.6(2)°) hydrogen bonds (Table S1). The *anti*-N–H of the **pic** forms an N–H···O (2.948(2) Å, 126.3(1)°) hydrogen bond with the same carboxylic oxygen atom of a symmetrically related acid molecule, and the hydroxyl O–H of the **D-H₂ma** forms an O–H···N (3.071(2) Å, 139.9(2)°) hydrogen bond with the adjacent pyridine N of the **pic**, forming a second heterosynthon of graph set $R_2^2(10)$. Thus, it generates a four-component supramolecular plane unit that consists of each two molecules of **D-H₂ma** and **pic**, giving rise to a new ring motif $R_4^2(8)$ in the same way as in the other polymorph [33] (Figure 1b).

Although little is known about the crystallization mechanisms involving the stages of molecular aggregation to form cocrystals [38], in this case it is likely that the mechanism is likely to include a first stage of molecular recognition between an acid molecule and another of amides, the well-known amide–acid heterosynthon. In a second stage, two of these heterodimers, symmetrically related, are associated by establishing new hydrogen bonds, to form the above tetrameters (Figure 1b).

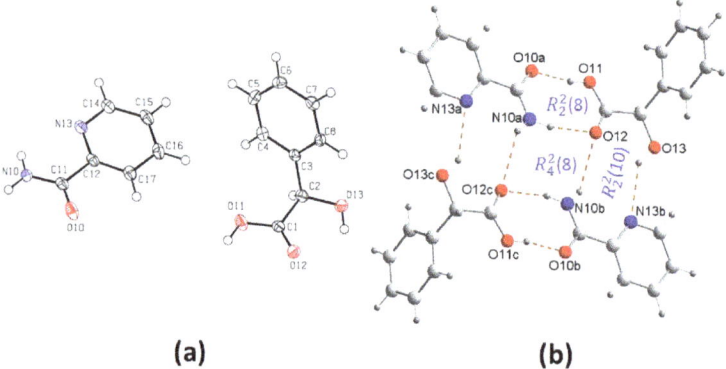

Figure 1. (a) A view of the molecular structure of **1** showing atomic labeling and displacement ellipsoids drawn at the 50% probability level, and (b) four-component supramolecular unit showing the intermolecular interactions and the supramolecular synthons. Hydrogen bonds are shown as orange dashed lines. See Table S1 for symmetry codes.

Another aspect to consider is the molecular conformation of each of the cocrystals and the difference between them and with mandelic acid and the conformers. Conformational flexibility of pyridinecarboxamides is related to the torsion angle of the amide group in relation to the pyridine ring (θ_1), and in mandelic acid to the torsion angles involving the carbonyl group of the carboxylic acid (θ_2) and the group hydroxyl of the alcohol group (θ_3) (Scheme 1). Table 2 compares the molecular conformations of pyridinecarboxamides and mandelic acid in these cocrystals.

Table 2. Torsional angles of pyridinecarboxamides and mandelic acid crystals and corresponding cocrystals.

Compound	Polymorph	θ_1 *	θ_2 *	θ_3 *	Molecule
pic	I	−161.3			
DL-H$_2$ma	I		−105.0	−65.8	
DL-H$_2$ma	II		−122.1	−91.5	1
	II		−125.7	−42.5	2
1		−168.5	−111.0	−25.8	
HOGGOB [31]		−173.3	125.5	90.9	1
		175.5	−79.2	54.3	2
nam	α	157.6			
	ε	26.8			
	ι	151.5			
D-H$_2$ma			−124.5	−80.5	1
			−122.0	−30.5	2
JILZOU01 (2a) [32]		−28.7	−106.1	−72.8	
JILZOU (2b) [33]		35.3	−129.6	−48.2	1
		12.0	−103.5	−41.6	2
inam	I	30.5			
	III	30.9			
	V	19.6			
L-H$_2$ma	I		119.4	22.9	1
			120.1	−87.4	2
	II		115.5	10.9	1
			116.8	100.0	2
3		−0.5	112.4	73.5	

* θ_1, θ_2 and θ_3 are defined in Scheme 1; values of torsional angles were calculated from the crystal structures in CSD.

In **1**, the oxygen atom of the amide group is opposite the nitrogen atom of the pyridine ring outside the pyridine plane, in a similar way to that of polymorph I of pure pic. In the mandelic acid molecule, small conformational changes are also observed, more pronounced in θ_3, which are consistent with those existing in the second symmetrically independent molecule of **D-H$_2$ma**, and close to those of the second molecule of polymorph II of **DL-H$_2$ma** (Table 2). When these values are compared with those of the HOGGOB polymorph, noticeable and more pronounced discrepancies are observed in the torsion angles of the second symmetrically independent acid molecule, not only in θ_3 but also in θ_2, probably due to the different hydrogen bonds in which the donor alcoholic OH groups participate.

In crystal packing, the flat units are arranged one next to the other in the plane "bc", without any interaction (Figure 2a), so that in the direction of the diagonal of the angle between the axes "a" and "c" are arranged intercalated in opposite orientations to form a 3D network with an internal zigzag-like orientation (Figure 2b). The absence of strong interactions between tetramers may justify for some softer properties in the cocrystals, compared to those of picolinamide.

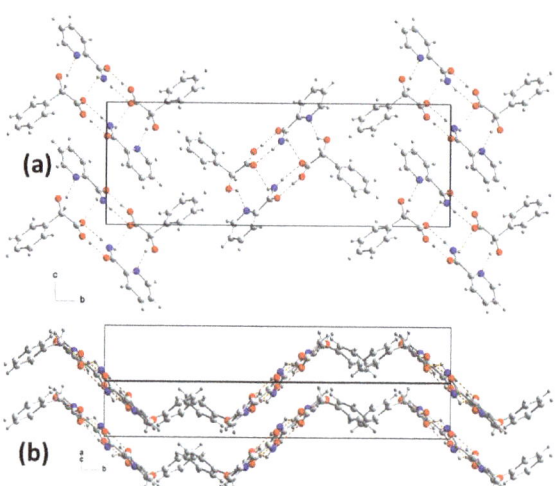

Figure 2. Packing diagram for **1**, viewed (**a**) in the "bc" plane and (**b**) parallel to "b" axis.

From the cocrystallization of **DL-H$_2$ma** and **nam** are known several structures of different stoichiometries in ratios 1:1, 2:1, 2:2 and 1:4 [23]. In this laboratory, the same (**nam**)-(**D-H$_2$ma**) (1:1) cocrystal was prepared by crystallization from ethyl acetate. Crystal structure analysis of **2** revealed that the cocrystal belongs to monoclinic, $P2_1$ space group with one molecule each of **D-H$_2$ma** and **nam** in the crystallographic asymmetric unit (Figure 3a). The crystal structure features an heterosynthon between the α-hydroxyl carbonyl group of **D-H$_2$ma** and the amide group of **nam** involving O–H···O (2.708(1) Å, 143.3(2)°) and N–H···O (3.002(1) Å, 155.4(2)°) with ring motif $R_2^2(9)$ (Table S1). These heterodimers are further joined by hydrogen bonds through O–H carboxyl acid and pyridine N atom (2.684(1) Å, 175.0(2)°), which is accompanied by a stabilizing C–H···O hydrogen bond (H···O, 2.342 Å; C···O, 3.103 Å), resulting in a supramolecular synthon of graph set $R_2^2(7)$. In addition, the amide *anti*-N–H and hydroxy O atom N–H···O (2.921(1) Å, 144.9(1)°) form heterosyntons of graph set $R_4^3(11)$, to originate a new four-component supramolecular unit that is repeated along infinite ribbons (Figure 3b). In the 3D network, these heterodimers are further joined by hydrogen bonds to form independent layers along "b" axis (Figure 4a) [32], extending in the "ca" plane simulating a zigzag chain that, unlike **1** (Figure 2b), all molecules are oriented in the same way (Figure 4b).

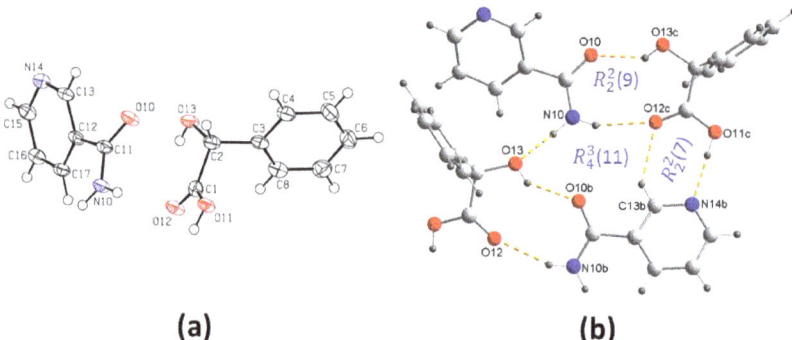

Figure 3. (**a**) A view of the molecular structure of **2** showing atomic labeling and displacement ellipsoids drawn at the 50% probability level, and (**b**) four-component supramolecular unit showing the hydrogen patterns, as orange dashed lines, observed in **2**. See Table S1 for symmetry codes.

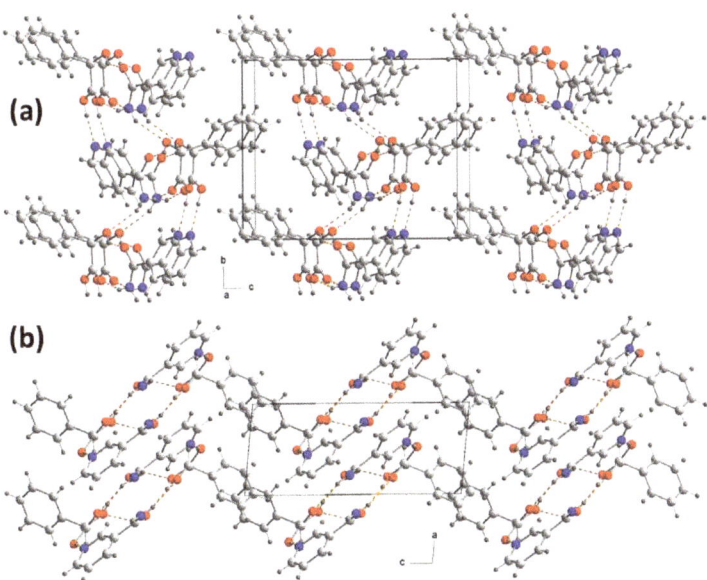

Figure 4. Packing diagram for **2**, viewed (**a**) in the "bc" plane and (**b**) in the "ca" plane.

In the **2** system, taking as reference the data of the structure of the polymorph ε (NICOAM07, from the CSD), in the crystals of pure **nam** it is observed that the O atom of the amide group is on the same side of the atom nitrogen of the pyridine ring (θ_1, 26.8°), which is the same conformation adopted by the polymorphs JILZOU01 [32] (−28.7°) and JILZOU [33] (35.3 and 12.0°). It should be noted that this conformation is opposite to that described above for the cocrystals of **1**. Regarding the **D-H$_2$ma** molecule in the cocrystal, differences in conformation are also observed, as can be seen in the values of θ_2 and θ_3 compared with those of the two symmetrically independent molecules of the acid, although they do not differ in excess of those found in HOGGOB molecule **1** [31] (Table 2).

The new cocrystals of **DL-H$_2$ma** with **inam** in a 1:1 ratio, prepared by crystallization in ethyl acetate, have also been previously obtained in warm ethanol [6]. The crystal structure revealed that cocrystals **3** belong to the monoclinic space group, $P2_1/c$ with one molecule of each coformer in the asymmetric crystal unit (Figure 5a). **L-H$_2$ma** and **inam** interact with each other via an acid–pyridine heterosynthon involving O–H···N (2.624 Å, 177.3°) hydrogen bond (Table S1). As in 2, the N_{py}···H−O hydrogen bond is accompanied by a complementary C−H···O hydrogen bond (H···O, 2.640 Å; C···O, 3.131 Å, CHO 127.9°). The amide group forms a amide–amide homosynthon dimers of typical ring motif $R_2^2(8)$ between **inam** molecules involving N–H···O (2.881(1) Å, 179.7(2)°). At the same time, these dimers are attached to **L-H$_2$ma** molecules in two ways. One is through a O–H···O bond formed between the hydroxyl O–H and the amide O of a nearest neighboring acid molecule (2.775(1) Å, 162.9(2)°) whereas the second one is via N–H···O (2.949(1) Å, 141.7(1)°) between the amine *anti*-N–H and the hydroxy oxygen of the **L-H$_2$ma** (Figure 5b), which is reinforced by a hydrogen bond C−H···O in which carbonyl participates, originating a heterosynthon of graph set $R_2^2(10)$ (Figure 5b).

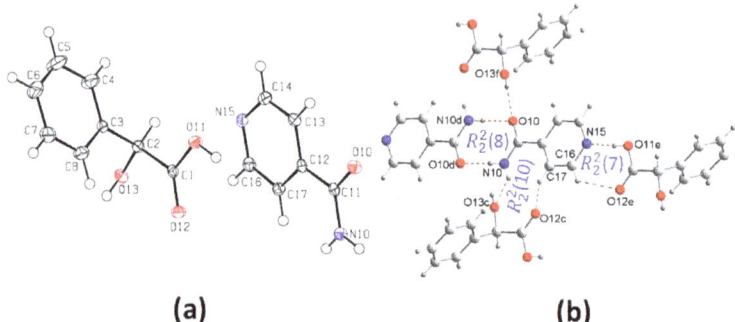

Figure 5. (a) A view of the molecular structure of **3** showing atomic labeling and displacement ellipsoids drawn at the 50% probability level, and (b) primary hydrogen-bond interactions in **3**. Hydrogen bonds are shown as orange dashed lines. See Table S1 for symmetry codes.

In **3**, **inam** is practically planar (θ_1, $-0.5°$), which contrasts with the values of this torsion angle found in polymorphs I, III and V of the pyridinecarboamide (Table 2). Conformational differences are also observed in the **L-H$_2$ma** molecule. Note especially that the four molecules of the two polymorphs of **L-H$_2$ma** have substantially different conformations from those found in the cocrystal, particularly θ_3.

Another way to describe the hydrogen bond interactions in the crystal packing is considering two self-complementary amide–amide pairwise homosynthon dimers between **inam** mutually parallel molecules linked through two molecules of mandelic acid each by hydrogen bonds of O–H···O type between OH of the alcohol group and the carbonyl O atom of each symmetrically related amide, giving rise to a supramolecular dimer of graph set $R_4^4(12)$, in the direction of the "a" axis (Figure 6a), and also in the plane "bc" by means of a carboxylic acid–pyridine interaction forming supramolecular heterosynthons of graph set $R_6^4(28)$ (Figure 6b). The set of these interactions gives rise to a crystal network constituted by independent layers parallel to the "a" axis (Figure 6c).

In the formation of cocrystals of **3**, the probable mechanism must include a first stage of molecular recognition between two amide molecules to form the well-known amide–amide homosynthon. In a second stage, two of those homodimers, symmetrically related, are associated by hydrogen bonds, to form the aforementioned tetramers (Figure 6a,b).

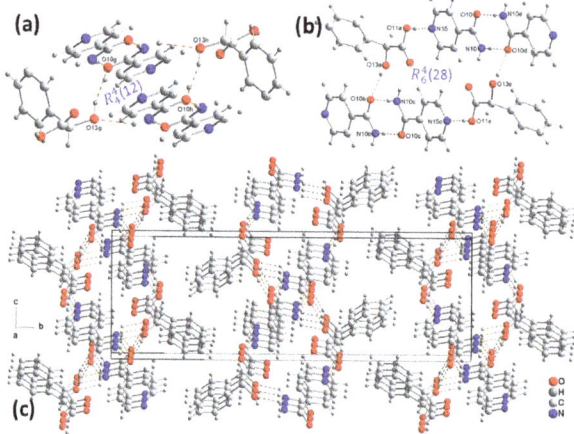

Figure 6. View of amide–amide homosynthon showing (a) O–H···O interactions, (b) O–H···N$_{py}$ interactions and (c) packing diagram for **3**, viewed in the "bc" plane.

3.2. Supramolecular Synthons

Supramolecular synthons are spatial arrangements of intermolecular interactions between complementary functional groups, and constitute the core of the retrosynthetic strategy for supramolecular structures [39]. With crystal structures defined as networks with the molecules as the nodes and the supramolecular synthons as the connections between them, retrosynthesis can be performed on network structures to produce appropriate molecular structures. The advantage of such an approach in crystal engineering is that comparisons between seemingly different crystal structures are facilitated. If we apply this observation to the systems studied here, we can establish some differences between them that can be attributed to the different constituent isomers.

With the structural data of the cocrystals studied here, we compare the main synthons that they present in their 3D network, which are displayed in Scheme 2. In cocrystal **1**, two heteromeric interactions are observed that give rise to two heterosynthons between *syn*-amide with carboxylic acid (B) and between pyridine amide with hydroxy carboxylic acid (E). In cocrystal **2**, two heteromeric interactions are also observed, causing two heterosynthons, one between pyridine and carboxylic acid (C) and the other between *syn*-amide and hydroxy carboxylic acid (D). In cocrystal **3**, on the contrary, a homomeric interaction corresponding to typical amide–amide homosinton (A) and two heteromeric interactions stand out, one is the heterosinton pyridine with carboxylic acid (C), also observed in the cocrystals of **2**, and the other *anti*-amide with hydroxy carboxylic acid (F).

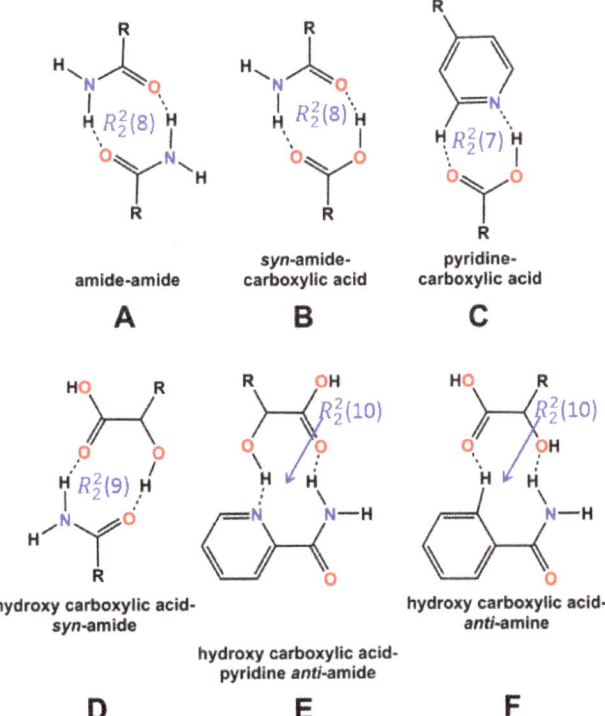

Scheme 2. Main supramolecular synthons in the cocrystals studied herein. Hydrogen bonds are shown as dotted lines.

3.3. Theoretical DFT Study

To analyze and understand the different modes of interaction observed in the solid state of the cocrystals, a density functional theory (DFT) study at the PBE0-D3/def2-

TZVP level of theory was carried out. The recurrent motifs observed in the solid-state X-ray structures are analyzed herein, focusing on the calculation of the individual H-bond energies and also the unconventional stacking interactions in compounds **1** and **3** that include the H-bonded arrays as described below.

Figure 7 shows the molecular electrostatic potential (MEP) surfaces of all coformers (pyridylcarboxamides and mandelic acid) studied in this work. The MEP surface analysis is useful to investigate the electron-rich and electron-poor regions of the crystal coformers. It can be observed that the most positive region corresponds to the H-atom of carboxylic acid (+58 kcal/mol) in mandelic acid followed by the H-atoms of the carboxamide groups. These are more positive in nicotinamide and isonicotinamide due to the influence of the aromatic H-atom in *ortho* (+52 and +53 kcal/mol for nicotinamide and isonicotinamide, respectively). Moreover, the most negative regions correspond to the O-atoms of the carboxamide group (ranging from −38 to −43 kcal/mol). The MEP values are also large and negative at the O-atom of the carboxy group in mandelic acid and the aromatic N-atom of nicotinamide and isonicotinamide rings (ranging from −30 to −35 kcal/mol). Finally, the MEP is slightly less negative at the aromatic N-atom of picolinamide (−17 kcal/mol) and hydroxyl group of mandelic acid (−27 kcal/mol). This analysis provides evidence that the carboxy group of mandelic acid is the strongest H-bond donor and the O-atom of the carboxamide group the best H-bond acceptor.

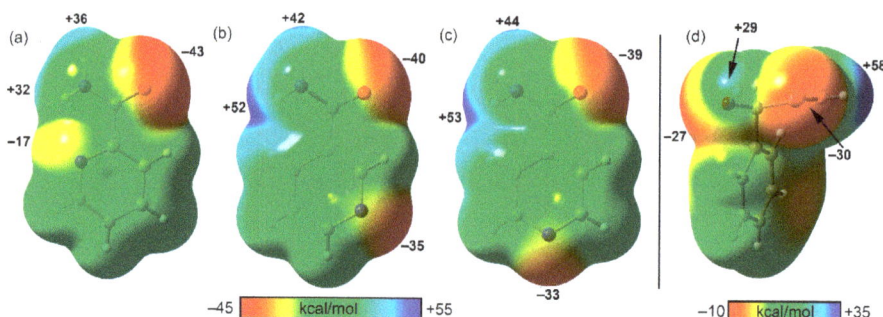

Figure 7. MEP surfaces of picolinamide (**a**), nicotinamide (**b**), isoniconitamide (**c**) and mandelic acid (**d**) at the PBE0-D3/def2-TZVP level of theory. Isosurfaces plotted using 0.001 a.u. The energies at selected points of the surface are indicated in kcal/mol.

The hydrogen bond and formation energies of the assemblies were estimated using the QTAIM method and the value of the Lagrangian kinetic G(r) contribution to the local energy density of electrons at the critical point (CP). The dissociation energy of each individual H-bonding interaction was estimated using the approach proposed by Vener et al. [40], which was specifically developed for HBs (Energy = 0.429 × G(r) at the bond CP).

Figure 8 represents the distribution of bond CPs and bond paths for the tetrameric assembly observed in the picolinamide–mandelic acid cocrystal exhibiting a network of H-bonds. The existence of a bond CP and bond path connecting two atoms is a universal indication of an interaction [41]. The QTAIM analysis of the tetrameric assembly represented in Figure 8a shows the presence of appropriate bond CPs (red spheres) and bond paths connecting the N/O-atoms to the H atoms in the H-bonding interactions. Moreover, several ring CPs (yellow spheres) also emerge upon complexation due to the formation of supramolecular rings. The distribution also shows the existence of weak C–H···O interactions between one aromatic C–H group and the hydroxyl O-atom of the mandelic acid. The dissociation energy of the tetrameric assembly is large (39.7 kcal/mol), thus confirming the importance of this H-bonding network. Figure 8b also includes the individual energy of each H-bond that is indicated in blue next to the bond CP that characterizes each H-bond. In agreement with the MEP analysis, the H-bonds involving the carboxy group as H-bond donor (O–H···O) are the strongest (11.0 and 7.8 kcal/mol). Moreover, several structure-guiding motifs are observed in the solid state structure of compound **1**. These are $R_2^2(8)$ and $R_4^4(8)$, involving only the carboxy and carboxamide groups, In addition, a $R_2^2(10)$ motif is also important, where the pyridine N-atom and the hydroxyl groups participate in addition to the dominant carboxy and carboxamide groups. We also evaluated energetically the formation of a different tetrameric assembly where two $R_2^2(8)$ motifs are stacked (see Figure 8c) in an antiparallel fashion. The binding energy computed as a dimerization energy of two $R_2^2(8)$ motifs is −9.7 kcal/mol, thus revealing the strong nature of this unconventional π-stacking interaction. The NCIplot index analysis is represented in Figure 8c, showing large and green (meaning attractive interaction) isosurfaces located between the carboxy and carboxamide groups of both crystal coformers, thus suggesting that the interaction involves the π-system of both groups ($\pi_{COOH}\cdots\pi_{CONH2}$). The NCI isosurface is dark blue for the OH···O and light blue for the NH···O H-bonds of the $R_2^2(8)$ motifs, thus confirming the strong nature of the OH···O bonds in line with the QTAIM dissociation energy and the MEP surface analysis.

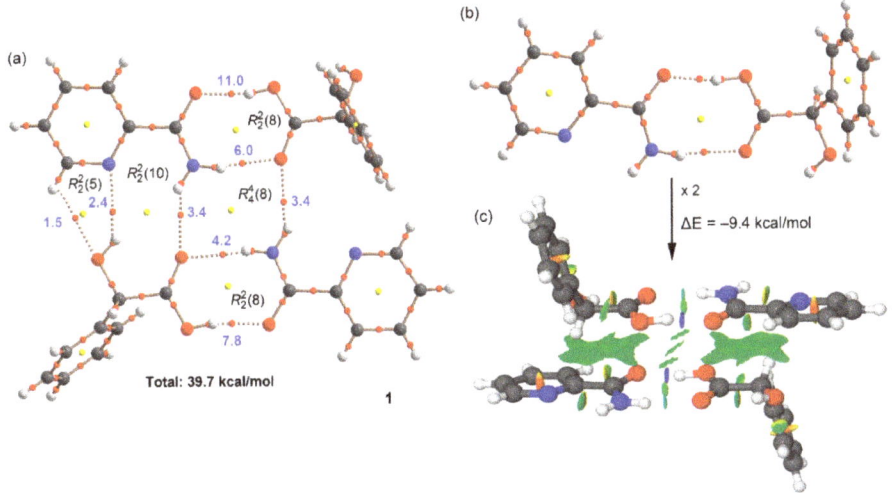

Figure 8. (a) AIM distribution of bond and ring critical points (green and yellow spheres, respectively) and bond paths obtained for the H-bond assembly of compound **1**. The dissociation energies of the H-bonds using the G(r) values are indicated next to the bond CPs. (b) AIM distribution of bond and ring critical points (green and yellow spheres, respectively) and bond paths obtained for the H-bonding $R_2^2(8)$ motif. (c) NCIplot index analysis of the π-stacked tetramer. |RGD| isosurface 0.5, density cutoff 0.04 a.u., color range: −0.04 a.u. ≤ (signλ$_2$) ρ ≤ 0.04 a.u.

For compounds **2a** (JILZOU01) and **2b** (JILZOU), a similar study was performed, where we selected a representative tetrameric assembly for each one including the most important interactions and motifs. The QTAIM analyses are shown in Figure 9, where each H-bond is characterized by a bond CP connecting the O/N-atoms to the H-atoms. Similar to **1**, the distribution shows the existence of weak C–H···O interactions between H-atoms of nicotinamide ring and the O-atoms of the hydroxyl groups in **2a** or carboxamide group in **2b**. Moreover, a recurrent $R_2^2(7)$ motif is observed in both compounds where the carboxy group of the mandelic acid forms a strong OH···N$_{py}$ H-bond (8.4 and 10.4 kcal/mol in **2a** and **2b**, respectively) combined with a much weaker C–H···O H-bond (2.6 and 1.7 kcal/mol in **2a** and **2b**, respectively). In **2a**, a $R_2^2(9)$ motif is observed where the hydroxyl group of mandelic acid establishes a moderately strong H-bond (5.7 kcal/mol) with the carboxamide group of the coformer. An interesting $R_2^2(8)$ motif is observed in **2b** involving the carboxamide groups (4.5 kcal/mol each H-bond) and leading to the formation of a self-assembled dimer. The formation energies of the selected tetramers are similar (24.6 kcal/mol in **2a** and 26.9 kcal/mol in **2b**).

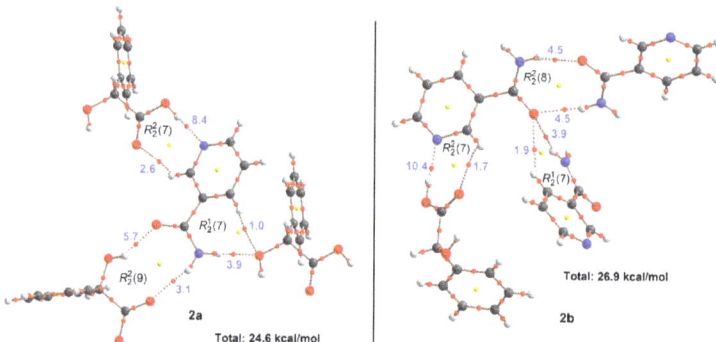

Figure 9. AIM distribution of bond and ring critical points (green and yellow spheres, respectively) and bond paths obtained for the H-bond tetrameric assemblies of compounds **2a** and **2b**. The dissociation energies of the H-bonds using the G(r) values are indicated next to the bond CPs.

In compound **3**, the tetrameric assembly shown in Figure 10a was analyzed, where the isonicotinamide molecule forms self-assembled dimers via two equivalent N–H···O bonds ($R_2^2(8)$ motif) with a total dissociation energy of 10.4 kcal/mol. In addition, the recurrent $R_2^2(8)$ motif described above in **2a** and **2b** is also observed in **3** with a dissociation energy of 13.8 kcal/mol, thus revealing that the combination of the strong OH···H and weak CH···O H-bonds is energetically favored over the two symmetric NH···O H-bonds between the carboxamide groups. This tetrameric assembly, which presents a very large dissociation energy (37.0 kcal/mol) self-assembly, forming π-stacking octamers in the solid state, as represented in Figure 10b. The NCIplot index analysis shows a much extended isosurface that embraces the whole assembly and explaining the large dimerization energy (−27.3 kcal/mol). The isosurfaces clearly show that the H-bonded arrays are also involved in the stacking interaction, as previously observed in the literature [42]. Actually, it has been demonstrated that the formation of H-bonded arrays is energetically enhanced over aromatic surfaces [43].

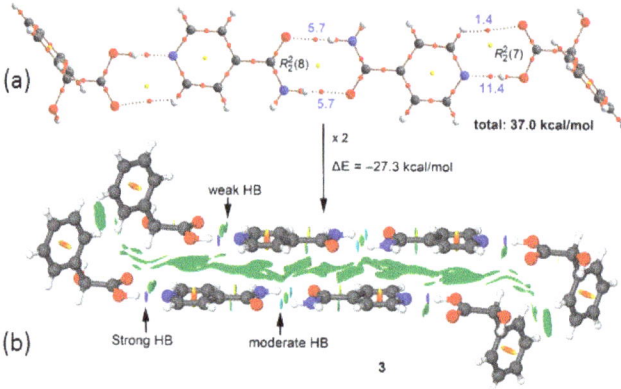

Figure 10. (a) AIM distribution of bond and ring critical points (green and yellow spheres, respectively) and bond paths obtained for the H-bonding assembly of compound **3**. The dissociation energies of the H-bonds using the G(r) values are indicated next to the bond CPs. (b) NCIplot index analysis of the π-stacked octamer. |RGD| isosurface 0.5, density cutoff 0.04 a.u., color range: −0.04 a.u. ≤ (signλ$_2$) ρ ≤ 0.04 a.u.

Table 3 summarizes the interaction energies for the H-bonding assemblies represented in Figures 8–10 computed using the supramolecular approach (BSSE corrected) and estimated using Verner's equation. In general, there is a good agreement between the BSSE corrected energies and those derived from QTAIM, giving reliability to Verner's approach. In some cases, such as the pentameric assembly of compound **1** (Figure 8a) and the tetramer of compound JILZOU (Figure 9b), the interaction energies are greater (in absolute value) than the formation energies derived from the QTAIM approach. This is due to an extra stabilization in those assemblies caused by van der Waals forces and other long-range interactions due to the approximation of the bulk of the molecules. In any case, the H-bonding interactions are clearly dominant.

Table 3. Interaction energies of the HB assemblies derived from the supramolecular approach (BSSE corrected) and derived from the QTAIM (E_{BSSE} and ΣE_{HB}, respectively) in kcal/mol at the PBE0-D3/def2-TZVP level of theory.

Compound	E_{BSSE}	ΣE_{HB}
(pic)-(**D-H$_2$ma**) (**1**) (Figure 8a)	−42.7	39.7
(pic)-(**D-H$_2$ma**) (**1**) (Figure 8b)	−17.0	16.9
JILZOU01 (**2a**) (Figure 9a)	−22.5	24.6
JILZOU (**2b**) (Figure 9b)	−32.3	26.9
(inam)-(**L-H$_2$ma**) (**3**) (Figure 10a)	−39.5	37.0

3.4. X-ray Powder Diffraction (XRPD) Analysis

The formation of cocrystals could be verified by XRPD. In Figure 11, each XRPD pattern of mandelic acid cocrystals is different from either that of **D, L-H$_2$ma** or the corresponding coformer. All of the peaks displayed in the measured XRPD patterns of the **D, L-H$_2$ma** cocrystal bulk powder are in close accordance with those in the simulated patterns acquired from single-crystal diffraction data, which confirm the formation of high-purity phases. Similarly, in both cocrystals, the XRPD patterns simulated from the single-crystal structures matches well with the XRPD patterns of powder obtained (Figure S12).

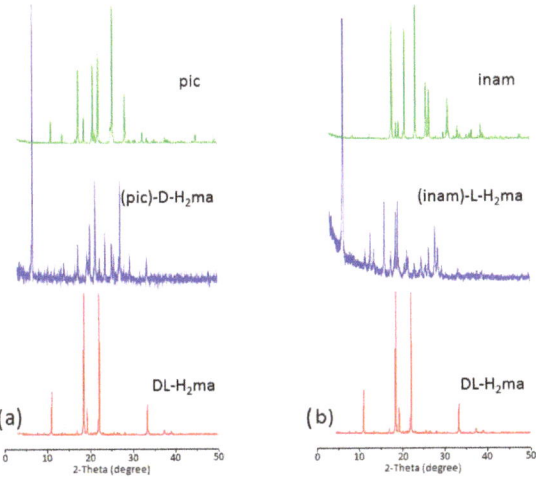

Figure 11. XRPD patterns of mandelic acid (red), pyridinecarboxamides (green) and cocrystal (blue) in (**a**) **1** and (**b**) **3**.

3.5. DSC Analysis

The thermal behavior of coformers and cocrystals was tested using differential scanning calorimetry (DSC). The DSC heating curves and melting temperatures are represented in Figure 12.

The melting point of the cocrystals, DL-Mandelic acid and pyridinecarboxamides are listed in Table 4. **H₂ma**, **pic**, **nam** and **inam** exhibit a sharp melting endotherm followed by decomposition, and do not show any phase change/transformation on heating before melting. Similarly, the (**pic**)-(**D-H₂ma**) (Figure 12a), (**nam**)-(**L-H₂ma**) (Figure 12b) and (**inam**)-(**L-H₂ma**) (Figure 12c) binary solids exhibit melting, followed by decomposition. There is no phase change before melting on heating. The cocrystals displayed a lower melting point than the pure mandelic acid and the coformers. The thermal analysis of the cocrystals and their comparison to that of the coformers revealed that in (**1**) and (**3**) there is a direct correlation between the melting points of the coformers and the cocrystals (Table 4): the higher the melting point of the coformer, the higher the melting point of the cocrystal. However, in (**nam**)-(**L-H₂ma**) this trend is broken and its melting point is the lowest of the three cocrystals, probably because the strength of the hydrogen bonds is also the weakest.

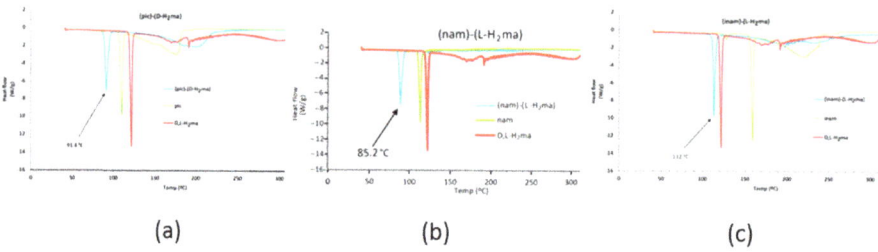

(a)   (b)   (c)

Figure 12. DSC analyses of the resulting powders of pure components and the cocrystals **1**, **2a** and **3**.

Table 4. Melting points (M. p.) of coformers * and prepared cocrystals.

Coformer	M. p. (°C)	Cocrystal	M. p. (°C)
DL-Mandelic acid (**D,L-H₂ma**)	118–121		
D/L-Mandelic acid (**L-H₂ma**)	131–135		
Picolinamide (**pic**)	104–108	(**pic**)-(**D-H₂ma**) (**1**)	91.4
Nicotinamide (**nam**)	128–131	(**nam**)-(**L-H₂ma**) (**2**) [25]	85.2
Isonicotinamide (**inam**)	155–157	(**inam**)-(**L-H₂ma**) (**3**)	112.0

* Taken from http://www.chemspider.com/.

3.6. Thermal Analysis

To determine the thermal stability of cocrystals, thermogravimetric analysis (TGA) was performed under a stream of air in the range 25–300 °C. Considering that the compounds are thermally stable until a 10% weight loss of the sample occurs, it can be seen that (**1**) is stable up to 179 °C, very similar to mandelic acid, while (**3**) undergoes a 10% weight loss when 209 °C is reached, so that its stability is comparable to that of isonicotinamide (Figure 13). Above these temperatures a slow weight loss is observed up to 300 °C, decomposing and releasing CO, CO_2 and various nitrogen oxides, not resulting in final solid waste. The phase purity of the as-synthesized samples could be confirmed by the PXRD pattern, where the characteristic peaks match well with those of the simulated PXRD pattern based on the single crystal data (Figure S10).

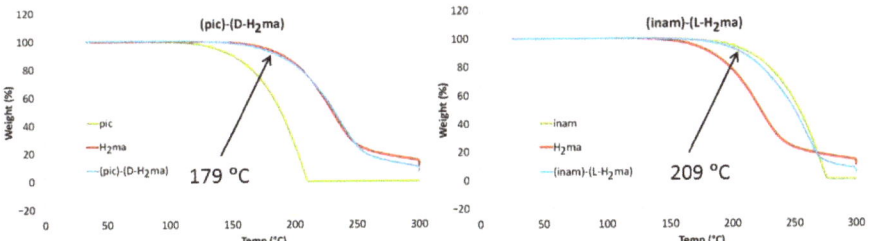

Figure 13. TG thermograms of pure components and the cocrystals **1** (**left**) and **3** (**right**).

3.7. FT–IR Spectroscopy

The IR spectra of cocrystals **1** and **3** are depicted in Figures S13 and S14, respectively. A comparison with the spectra of its coformers is represented in Figure 14. Regions 3500–3000 and 1700–600 cm^{-1} are important sources of information about the molecular interactions, because these two regions are associated with hydrogen bonding interaction. In solid DL-mandelic acid, these bands correspond to 3397 cm^{-1}, ν(OH), 1716 cm^{-1}, ν(C=O) and 1078–1064 cm^{-1}, ν(C-O) [44], while that in pyridine carboxamides appear at 3414 and 3164 cm^{-1}, ν(NH$_2$), 1658 cm^{-1}, ν(C=O), 1603 cm^{-1}, δ(NH$_2$), 1386 cm^{-1}, ν(CN) and 640–629 cm^{-1}, α(CCC), in picolinamide; and 3362 and 3178 cm^{-1}, ν(NH$_2$), 1655 cm^{-1}, ν(C=O), 1622 cm^{-1}, δ(NH$_2$), 1390 cm^{-1}, ν(CN) and 668–614 cm^{-1}1, α(CCC), in isonicotinamide [45]. Consequently, in cocrystals, the absorb peaks around 3400 and 3200 cm^{-1} attribute to stretching vibrations of OH and NH$_2$ groups of the acid and pyridine carboxamides, where the wavenumbers for the asymmetric stretching vibration of NH$_2$ in cocrystal **1** are about 3435 cm^{-1}, while cocrystal **3** shows absorb peaks at about 3379 cm^{-1} due to stronger hydrogen bond interactions than in **1**. In the ν(NH$_2$) symmetric stretching vibrations, the same behavior is observed for **3** respective to **1**. For the two cocrystals, the wavenumbers at around 3450 cm^{-1} indicate the O−H···O hydrogen bond interactions between them, while the slight differences of the wavenumbers for NH$_2$ groups may attribute to the different hydrogen bond interaction experienced by it. While for the region around 1700–600 cm^{-1}, which are corresponding to C=O group, the almost identical plots for the two cocrystals, at 1700–1600 cm^{-1}, indicate similar hydrogen bond interactions around the C=O groups for **pic**, and **inam** in cocrystals **1** and **3**, respectively, but in **3** the band of mean intensity at 1729 cm^{-1} corresponds to the C=O of **H$_2$ma** that does not take part in hydrogen bonding, which are in accordance with the structural analysis. We also point out the presence of two new bands at 2506 and 1913 cm^{-1} for **1** and 2465 and 1891 cm^{-1} for **3** that result from the O−H···N$_{py}$ hydrogen bond. This provides clear proof that the hydroxyl or carboxylic group of **H$_2$ma** interacts with the aromatic nitrogen of pyridine carboxamides [46]. A similar behavior is deduced when observing the position of in-plane and out-of-plane ring deformation bands of pyridinecarboxamides in the cocrystals, which reflect a greater strength of the O−H···N interaction in **3** than in **1** (Table S1), since while in **1** it is O−H$_{(hydroxyl)}$···N$_{(py)}$ with a distance O···N of 3.071 Å, in **3** is 2.624 Å for that distance in O−H$_{(carboxylic)}$···N$_{(py)}$.

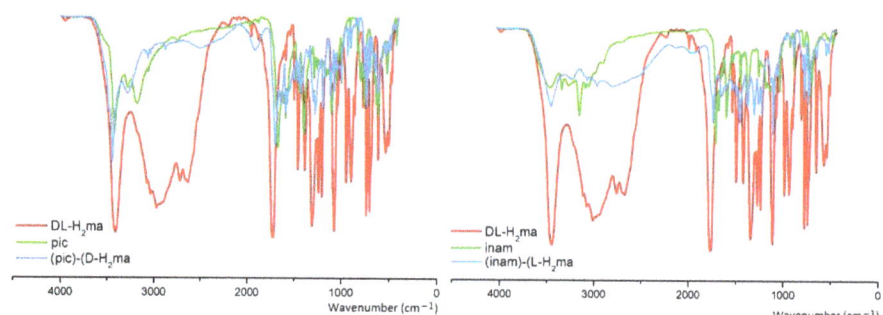

Figure 14. IR spectra of the cocrystals **1** (**left**) and **3** (**right**).

3.8. Solubility and Dissolution

Drug solubility occurs under a dynamic equilibrium state, which determines the maximum concentration in a saturated solution when excess solid is present in the media. The dissolution rate is a kinetic process that measures the drug concentration, which passes into the media with respect to time. These two parameters are determined by the solvation of the molecular components and the strength of the crystal structure lattice [47]. To improve the drug solubility, the solvent affinity must be increased and/or lower the lattice energy. These can be altered via cocrystal formation, although it is also influenced by the coformer solubility [48]. Since the aqueous solubility values pyridinecarboxyamides are greater than those of **D, L-H₂ma**, the presence of coformers may appear to improve the dissolution profile of mandelic acid (Figure 15).

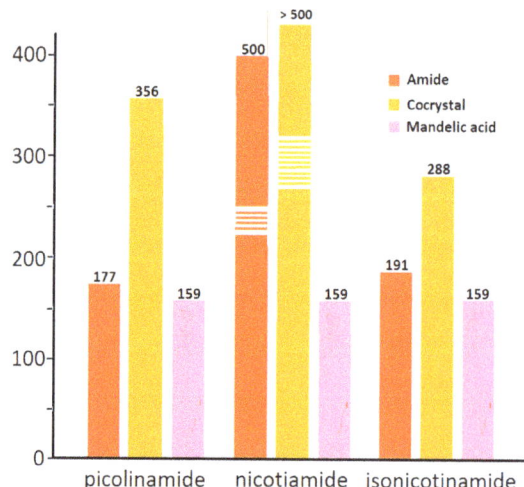

Figure 15. Solubility values in water (mg mL^{-1}) of the coformers and cocrystals **1**–**3**.

4. Concluding Remarks

The synthesis and X-ray characterization of two new cocrystals of mandelic acid and pyridylcarboxamides are reported herein. There is competition and interplay of the hydrogen bonding functional groups during binary cocrystallization. The results observed suggest that the hydroxy carboxylic acid forms reliable synthons to afford cocrystals with the pyridinecarboxamides. For the formation of tetramers in **1** and **3**, a two-stage-based crystallization mechanism is proposed. The first of acid–amide or amide–amide molecular recognition, respectively, and the second per association, by hydrogen bond, heterodimers or homodimers, are symmetrically related. The energetic features of the H-bonds were studied using the QTAIM and MEP surface analyses evidencing that the COOH···N, O H-bonds are the strongest. Some recurrent motifs, such as $R_2^2(7)$ between carboxy group and the pyridine ring and the self-assembled $R_2^2(8)$ motif between the carboxamide groups, are described and analyzed energetically. We believe that the estimation of individual contributions by means of QTAIM analysis reported herein is useful in terms of rationalizing the interactions and for future design and synthesis of cocrystals.

Supplementary Materials: The following supporting information can be downloaded at: https://www.mdpi.com/article/10.3390/cryst12020142/s1, Figure S1: ^{1}H NMR spectra in DMSO-d$_6$ of cocrystals **1** (upper) and **3** (lower) comparative with their coformers; Figure S2: ^{1}H NMR spectrum of **DL-H$_2$ma**; Figure S3: ^{1}H NMR spectrum of **pic**; Figure S4. ^{1}H NMR spectrum of (**pic**)-(**D-H$_2$ma**) (**1**); Figure S5: ^{1}H NMR spectrum of **inam**; Figure S6: ^{1}H NMR spectrum of (**inam**)-(**L-H$_2$ma**) (**3**); Figure S7: ^{13}C NMR spectrum of DL-H$_2$ma; Figure S8: ^{13}C NMR spectrum of **pic**; Figure S9: ^{13}C NMR spectrum of (**pic**)-(**D-H$_2$ma**) (**1**); Figure S10: ^{13}C NMR spectrum of **inam**; Figure S11: ^{13}C NMR spectrum of (**inam**)-(**L-H$_2$ma**) (**3**); Figure S12: XRPD patterns of the solid forms of **1** and **3**, obtained at room temperature. The XRPD patterns of the cocrystals match well with the simulated XRPD. Table S1: Hydrogen bond parameters [Å, °] for cocrystals. Letters included as superscripts refer to symmetry codes shown in text and figures; Figure S13: IR spectrum of (**pic**)-(**D-H$_2$ma**) (**1**); Figure S14: IR Spectrum of (**inam**)-(**L-H$_2$ma**) (**3**).

Author Contributions: Conceptualization, I.G.-S., R.T.-I. and J.M.G.-P.; methodology, all authors; software, A.F.; validation, A.C., A.F. and J.N.-G.; formal analysis, I.G.-S., R.T.-I. and J.M.G.-P.; investigation, R.T.-I. and J.M.G.-P.; writing—original draft preparation, A.C. and A.F.; writing—review and editing, all authors; visualization, I.G.-S., R.T.-I. and J.M.G.-P.; supervision, A.C., J.N.-G. and A.F.; project administration, A.C., J.N.-G. and A.F.; funding acquisition, A.C. and A.F. All authors have read and agreed to the published version of the manuscript.

Funding: The study received no external funding.

Institutional Review Board Statement: Not applicable.

Informed Consent Statement: Not applicable.

Data Availability Statement: Not applicable.

Acknowledgments: We thank the research groups GI-1580 (USC, Xunta de Galicia) and FQM-283 (Junta de Andalucía) and the "Centre de Tecnologies de la Informació" (Universitat de les Illes Balears, UIB) for free allocation of computer time.

Conflicts of Interest: The authors declare no conflict of interest.

References

1. Nangia, A.K.; Desiraju, G.R. Crystal Engineering: An Outlook for the Future. *Angew. Chem. Int. Ed.* **2019**, *58*, 4100–4107. [CrossRef] [PubMed]
2. Desiraju, G.R.; Vittal, J.J.; Ramanan, A. *Crystal Engineering: A Text Book*; World Scientific: Singapore, 2011.
3. Desiraju, G.R. Crystal Engineering: From Molecule to Crystal. *J. Am. Chem. Soc.* **2013**, *135*, 9952–9967. [CrossRef]
4. Karimi-Jafari, M.; Padrela, L.; Walker, G.M.; Croker, D.M. Creating Cocrystals: A Review of Pharmaceutical Cocrystal Preparation Routes and Applications. *Cryst. Growth Des.* **2018**, *18*, 6370–6387. [CrossRef]
5. Solomos, M.A.; Mohammadi, C.; Urbelis, J.H.; Koch, E.S.; Osborne, R.; Usala, C.C.; Swift, J.A. Predicting Cocrystallization Based on Heterodimer Energies: The Case of N,N′-Diphenylureas and Triphenylphosphine Oxide. *Cryst. Growth Des.* **2015**, *15*, 5068–5074. [CrossRef]
6. Solomos, M.A.; Watts, T.A.; Swift, J.A. Predicting Cocrystallization Based on Heterodimer Energies: Part II. *Cryst. Growth Des.* **2017**, *17*, 5073–5079. [CrossRef]
7. Berry, D.J.; Seaton, C.C.; Clegg, W.; Harrington, R.W.; Coles, S.J.; Horton, P.N.; Hursthouse, M.B.; Storey, R.; Jones, W.; Friščić, T.; et al. Applying Hot-Stage Microscopy to Co-Crystal Screening: A Study of Nicotinamide with Seven Active Pharmaceutical Ingredients. *Cryst. Growth Des.* **2008**, *8*, 1697–1712. [CrossRef]
8. Aakeröy, C.B.; Beatty, A.M.; Helfrich, B.A. A High-Yielding Supramolecular Reaction. *J. Am. Chem. Soc.* **2002**, *124*, 14425–14432. [CrossRef] [PubMed]

9. Sarma, B.; Reddy, L.S.; Nangia, A. The Role of π-Stacking in the Composition of Phloroglucinol and Phenazine Cocrystals. *Cryst. Growth Des.* **2008**, *8*, 4546–4552. [CrossRef]
10. Shattock, T.R.; Arora, K.K.; Vishweshwar, P.; Zaworotko, M.J. Hierarchy of Supramolecular Synthons: Persistent Carboxylic Acid · · · Pyridine Hydrogen Bonds in Cocrystals That also Contain a Hydroxyl Moiety. *Cryst. Growth Des.* **2008**, *8*, 4533–4545. [CrossRef]
11. Borba, A.; Gómez-Zavaglia, A.; Fausto, R. Molecular Structure, Vibrational Spectra, Quantum Chemical Calculations and Photochemistry of Picolinamide and Isonicotinamide Isolated in Cryogenic Inert Matrixes and in the Neat Low-Temperature Solid Phases. *J. Phys. Chem. A* **2008**, *112*, 45–57. [CrossRef]
12. Olsen, R.A.; Liu, L.; Ghaderi, N.; Johns, A.; Hatcher, M.E.; Mueller, L.J. The Amide Rotational Barriers in Picolinamide and Nicotinamide: NMR and ab Initio Studies. *J. Am. Chem. Soc.* **2003**, *125*, 10125–10132. [CrossRef] [PubMed]
13. Alvarez-Lorenzo, C.; Castiñeiras, A.; Frontera, A.; García-Santos, I.; González-Pérez, J.M.; Niclós-Gutiérrez, J.; Rodríguez-González, I.; Vílchez-Rodríguez, E.; Zaręba, J.K. Recurrent motifs in pharmaceutical cocrystals involving glycolic acid: X-ray characterization, Hirshfeld surface analysis and DFT calculations. *Cryst. Eng. Comm.* **2020**, *22*, 6674–6689. [CrossRef]
14. Évora, A.O.L.; Castro, R.A.E.; Maria, T.M.R.; Rosado, M.T.S.; Ramos Silva, M.; Canotilho, J.; Eusébio, M.E.S. Resolved structures of two picolinamide polymorphs. Investigation of the dimorphic system behaviour under conditions relevant to co-crystal synthesis. *Cryst. Eng. Comm.* **2012**, *14*, 8649–8657. [CrossRef]
15. Li, X.; Ou, X.; Wang, B.; Rong, H.; Wang, B.; Chang, C.; Shi, B.; Yu, L.; Lu, M. Rich polymorphism in nicotinamide revealed by melt crystallization and crystal structure prediction. *Commun. Chem.* **2020**, *3*, 152–160. [CrossRef]
16. Aakeröy, C.B.; Beatty, A.M.; Helfrich, B.A.; Nieuwenhuyzen, M. Do Polymorphic Compounds Make Good Cocrystallizing Agents? A Structural Case Study that Demonstrates the Importance of Synthon Flexibility. *Cryst. Growth Des.* **2003**, *3*, 159–165. [CrossRef]
17. Li, J.; Bourne, S.A.; Caira, M.R. New polymorphs of isonicotinamide and nicotinamide. *Chem. Commun.* **2011**, *47*, 1530–1532. [CrossRef]
18. Eccles, K.S.; Deasy, R.E.; Fábián, L.; Braun, D.E.; Maguired, A.R.; Lawrence, S.E. Expanding the crystal landscape of isonicotinamide: Concomitant polymorphism and co-crystallisation. *Cryst. Eng. Comm.* **2011**, *13*, 6923–6925. [CrossRef]
19. Vicatos, A.I.; Caira, M.R. A new polymorph of the common coformer isonicotinamide. *Cryst. Eng. Comm.* **2019**, *21*, 843–849. [CrossRef]
20. Van Putten, P.L. Mandelic acid and urinary tract infections. *Ant. Van Leeuwenhoek* **1979**, *45*, 622–623. [CrossRef]
21. Cai, W.; Marciniak, J.; Andrzejewski, M.; Katrusiak, A. Pressure Effect on D,L-Mandelic Acid Racemate Crystallization. *J. Phys. Chem. C* **2013**, *117*, 7279–7285. [CrossRef]
22. Marciniak, J.; Andrzejewski, M.; Cai, W.; Katrusiak, A. Wallach's Rule Enforced by Pressure in Mandelic Acid. *J. Phys. Chem. C* **2014**, *118*, 4309–4313. [CrossRef]
23. Zhang, S.-W.; Harasimowicz, M.T.; de Villiers, M.M.; Yu, L. Cocrystals of Nicotinamide and (R)-Mandelic Acid in Many Ratios with Anomalous Formation Properties. *J. Am. Chem. Soc.* **2013**, *135*, 18981–18989. [CrossRef]
24. Young, D.; Ding, F.; Lipparini, F.; Egidi, F.; Goings, J.; Peng, B.; Petrone, A.; Henderson, T.; Ranasinghe, D.; Zakrzewski, V.G.; et al. *Gaussian 16, Revision A.01*; Gaussian, Inc.: Wallingford, UK, 2016.
25. Weigend, F.; Ahlrichs, R. Balanced basis sets of split valence, triple zeta valence and quadruple zeta valence quality for H to Rn: Design and assessment of accuracy. *Phys. Chem. Chem. Phys.* **2005**, *7*, 3297–3305. [CrossRef] [PubMed]
26. Weigend, F. Accurate Coulomb-fitting basis sets for H to Rn. *Phys. Chem. Chem. Phys.* **2006**, *8*, 1057–1065. [CrossRef]
27. Adamo, C.; Barone, V. Toward reliable density functional methods without adjustable parameters: The PBE0 model. *J. Chem. Phys.* **1999**, *110*, 6158–6169. [CrossRef]
28. Boys, S.F.; Bernardi, F. The calculation of small molecular interactions by the differences of separate total energies. Some procedures with reduced errors. *Mol. Phys.* **1970**, *19*, 553–566. [CrossRef]
29. Contreras-García, J.; Johnson, E.R.; Keinan, S.; Chaudret, R.; Piquemal, J.-P.; Beratan, D.N.; Yang, W. NCIPLOT: A Program for Plotting Noncovalent Interaction Regions. *J. Chem. Theory Comput.* **2011**, *7*, 625–632. [CrossRef] [PubMed]
30. Keith, T.A. *AIMAll (Version 13.05.06)*; TK Gristmill Software: Overland Park, KS, USA, 2013.
31. Chan, H.C.S.; Woollam, G.R.; Wagner, T.; Schmidtc, M.U.; Lewis, R.A. Can picolinamide be a promising cocrystal former? *Cryst. Eng. Comm.* **2014**, *16*, 4365–4368. [CrossRef]
32. Zhang, S.-W.; Guzei, I.A.; de Villiers, M.M.; Yu, L.; Krzyzaniak, J.F. Formation Enthalpies and Polymorphs of Nicotinamide−R-Mandelic Acid Co-Crystals. *Cryst. Growth Des.* **2012**, *12*, 4090–4097. [CrossRef]
33. Friščić, T.; Jones, W. Cocrystal architecture and properties: Design and building of chiral and racemic structures by solid–solid reactions. *Faraday Discuss* **2007**, *136*, 167–178. [CrossRef] [PubMed]
34. Gamidi, R.K.; Rasmuson, Å.C. Estimation of Melting Temperature of Molecular Cocrystals Using Artificial Neural Network Model. *Cryst. Growth Des.* **2017**, *17*, 175–182. [CrossRef]
35. Kerr, H.E.; Softley, L.K.; Suresh, K.; Hodgkinsona, P.; Evans, I.R. Structure and physicochemical characterization of a naproxen–picolinamide cocrystal. *Acta Crystallogr. Sect. C Cryst. Struct. Commun.* **2017**, *73*, 168–175. [CrossRef] [PubMed]
36. Bhogala, B.R.; Basavoju, S.; Nangia, A. Tape and layer structures in cocrystals of some di- and tricarboxylic acids with 4,49-bipyridines and isonicotinamide. From binary to ternary cocrystals. *Cryst. Eng. Comm.* **2005**, *7*, 551–562. [CrossRef]
37. Cruz-Cabeza, A.J. Acid–base crystalline complexes and the pK_a rule. *Cryst. Eng. Comm.* **2012**, *14*, 6362–6365. [CrossRef]

38. Lopes, L.C.; Orlando, T.; Simões, P.H.B.; Farias, F.F.S.; Bonacorso, H.G.; Zanatta, N.; Salbego, P.R.S.; Martins, M.A.P. Persistence of N-H···O=C Interactions in the Crystallization Mechanisms of Trisubstituted Bis-Ureas with Bulky Substituents. *Cryst. Growth Des.* **2021**, *21*, 5740–5751. [CrossRef]
39. Desiraju, G.R.; Steiner, T. *The Weak Hydrogen Bond in Structural Chemistry and Biology*; Oxford University Press: Oxford, UK, 2001.
40. Vener, M.V.; Egorova, A.N.; Churakov, A.V.; Tsirelson, V.G. Intermolecular hydrogen bond energies in crystals evaluated using electron density properties: DFT computations with periodic boundary conditions. *J. Comput. Chem.* **2012**, *33*, 2303–2309. [CrossRef]
41. Bader, R.F.W. A Bond Path: A Universal Indicator of Bonded Interactions. *J. Phys. Chem. A* **1998**, *102*, 7314–7323. [CrossRef]
42. Guo, D. Sijbesma, R.P.; Zuilhof, H. π-Stacked Quadruply Hydrogen-Bonded Dimers: π-Stacking Influences H-Bonding. *Org. Lett.* **2004**, *6*, 3667–3670. [CrossRef]
43. Bhattacharyya, M.K.; Saha, U.; Dutta, D.; Frontera, A.; Verma, A.K.; Sharma, P.; Das, A. Unconventional DNA-relevant π-stacked hydrogen bonded arrays involving supramolecular guest benzoate dimers and cooperative anion–π/π–π/π–anion contacts in coordination compounds of Co(ii) and Zn(ii) phenanthroline: Experimental and theoretical studies. *New J. Chem.* **2020**, *44*, 4504–4518. [CrossRef]
44. Badawi, H.M.; Förner, W. Analysis of the infrared and Raman spectra of phenylacetic acid and mandelic (2-hydroxy-2-phenylacetic) acid. *Spectrochim. Acta Part A* **2011**, *78*, 1162–1167. [CrossRef]
45. Bakiler, M.; Bolukbasi, O.; Yilmaz, A. An experimental and theoretical study of vibrational spectra of picolinamide, nicotinamide, and isonicotinamide. *J. Mol. Struct.* **2007**, *826*, 6–16. [CrossRef]
46. Castro, R.A.E.; Ribeiro, J.D.B.; Maria, T.M.R.; Ramos Silva, M.; Yuste-Vivas, C.; Canotilho, J.; Eusébio, M.E.S. Naproxen Cocrystals with Pyridinecarboxamide Isomers. *Cryst. Growth Des.* **2011**, *11*, 5396–5404. [CrossRef]
47. Suresh, K.; Minkov, V.S.; Namila, K.K.; Derevyannikova, E.; Losev, E.; Nangia, A.; Boldyreva, E.V. Novel Synthons in Sulfamethizole Cocrystals: Structure−Property Relations and Solubility. *Cryst. Growth Des.* **2015**, *15*, 3498–3510. [CrossRef]
48. Mannava, M.K.C.; Gunnam, A.; Lodagekar, A.; Shastri, N.R.; Nangia, A.K.; Solomon, K.A. Enhanced solubility, permeability, and tabletability of nicorandil by salt and cocrystal formation. *CrystEngComm* **2021**, *23*, 227–237. [CrossRef]

Article

Salification Controls the In-Vitro Release of Theophylline

Laura Baraldi [1,2], Luca Fornasari [2], Irene Bassanetti [2], Francesco Amadei [2], Alessia Bacchi [1,3] and Luciano Marchiò [1,3,*]

1. Dipartimento di Scienze Chimiche, della Vita e della Sostenibilità Ambientale, Università di Parma, 43124 Parma, Italy; laura.baraldi@unipr.it (L.B.); alessia.bacchi@unipr.it (A.B.)
2. Preclinical Analytics and Early Formulations Department, Chiesi Farmaceutici S.p.A., Largo Belloli, 43123 Parma, Italy; l.fornasari@chiesi.com (L.F.); i.bassanetti@chiesi.com (I.B.); f.amadei@chiesi.com (F.A.)
3. Biopharmanet TEC, Tecnopolo Padiglione 33, Università di Parma, 43124 Parma, Italy
* Correspondence: luciano.marchio@unipr.it

Abstract: Sustained released formulation is the most used strategy to control the efficacy and the adverse reactions of an API (active pharmaceutical ingredient) with a narrow therapeutic index. In this work, we used a different way to tailor the solubility and diffusion of a drug. Salification of Theophylline with Squaric Acid was carried out to better control the absorption of Theophylline after administration. Salification proved to be a winning strategy decreasing the dissolution of the APIs up to 54% with respect to Theophylline. Most importantly, this was accomplished in the first 10 min of the dissolution process, which are the most important for the API administration. Two polymorphs were identified and fully characterized. Theophylline squarate was discovered as trihydrate (SC-XRD) and as a metastable anhydrous form. Indeed, during the Variable Temperature-XRPD experiment, the trihydrate form turned back into the two starting components after losing the three molecules of water. On the other hand, the synthesis of the trihydrate form was observed when a simple mixing of the two starting components were exposed to a high humidity relative percentage (90% RH).

Keywords: xanthines; theophylline; squaric acid; controlled release; dissolution; solubility

1. Introduction

Xanthines like theophylline, caffeine, and theobromine are a group of alkaloids, which act as mild stimulants and as bronchodilators. Theophylline in particular has an anti-inflammatory effect in asthma and chronic obstructive pulmonary disease (COPD) at lower concentrations [1].

Theophylline's mechanism of action implies the inhibition of phosphodiesterase, which is responsible for the smooth muscle relaxation. However, theophylline is also known for its narrow therapeutic index (30–100 µM) caused by a remarkably low selectivity. Indeed, a concentration of 110 µM already leads to a wide range of adverse reactions such as nausea, vomiting, metabolic acidosis, and arrhythmias [2]. For this reason, despite being an effective API, theophylline is not considered the first choice in the treatment of asthma. It is usually administered orally in slow-release preparations for chronic treatment in combination with a short acting β2-agonist, long acting β2 agonist, or an inhaled corticosteroid [3–5].

Lots of significant work has been done in the formulation field to optimize the theophylline absorption profile. Pezoa et al. [6], Rodrigues et al. [7], and Jian et al. [8] are just a few examples of the huge effort that has been done so far aiming at a better in-vivo performance. Formulations were modified using excipients like Eudragit or involving nanoparticles to have a prolonged effect and better control on the adverse reactions [9]. While optimizing a formulation many aspects must be taken into account [10], such as the role of every single component and how they influence the final release or the scale up. The interactions between excipient-excipient and drug-excipient are also to be considered.

Hydrolysis of the drug, ion interactions leading to new insoluble forms, or other physical interactions like adsorption of the API onto the surface of excipients might be significant drawbacks, resulting from the interactions between excipients and drugs [11].

In this work, we used a different strategy to tune the properties of this active pharmaceutical ingredient: a salification of theophylline with squaric acid was carried out to optimize its solubility, dissolution rate, and diffusion. Squaric acid is a strong acid belonging to the family of oxocarbons [12]. It has been studied extensively for its peculiar physical properties [13] and for its use in the synthesis of dyes [14,15]. In the pharmaceutical industry, squaric acid derivatives have attracted interest for their potential use as topical immunosensitizers [16], enzyme inhibitors, and receptor antagonists [17]. However, to the best of our knowledge, squaric acid has never been studied in the formation of API salts. From a crystal engineering perspective, the features that guided the selection of this acid as salt-former are its strong acidity (pK_{a1} = 0.6, pK_{a2} = 3.4) [18], its highly symmetric character, and the well-established tendency to form layered H-bonded assemblies [19,20], that should couple effectively with mostly planar molecules such as xanthines.

Usually, when looking for a new salt to improve the pharmacokinetic properties of an API, hydrate forms are unwelcome. Hydrates usually have a higher thermodynamic stability leading to a slower dissolution profile [21,22]. In this peculiar case, the hydrate form would turn useful, providing a better control of the released drug. Moreover, salification might turn to be easier and quicker with respect to the formulation development. Indeed, once the salt is synthesized already known manufacturing processes can be applied, yielding standard tablets or capsules for instance. Some theophylline salts have already been studied [23,24]. Furthermore, some systems have been designed with the aim of improving bioactivity [25] or stability [26], however, in both cases, the salts resulted more soluble than theophylline alone. Hence, salification has never been applied to theophylline with the purpose to ameliorate the therapeutic index issues

2. Materials and Methods

2.1. General

Theophylline, Squaric acid, Tetrahydrofurane (THF), Diethyl Ether, and Ethanol were used as received from MERCK-Sigma Aldrich (Taufkirchen, Germany, EU). Solvents were commercially available and used without any other purification. Water was used after filtration with MilliQ Advantage A10 technology from Millipore (MERCK-Sigma Aldrich (Darmstadt, Germany)).

2.2. Theophylline Squarate

The bulk powder of the hydrate form (**TS3w**) was obtained via LAG (liquid assisted grinding). Theophylline and Squaric Acid were weighed in an equimolar ratio (1:1, total mass 163 mg), the sample was transferred to a 4.0 mL vial with three grinding balls (5 mm ϕ, zirconia type) and the desired solvent was added (20 μL of water Milli-Q). Then, the mixture was ground for 1 h using a multi-sample vibrating ball mill (Pulverisette 6—Fritsch, Germany) under a rotational speed of 300 rpm and a temperature of 25 °C.

Along with **TS3w**, an anhydrous form was obtained (**TSan**). In order to obtain **TSan**, Theophylline (100 mg) and Squaric Acid (63 mg) were both dissolved in water (10 ml) at room temperature and lyophilized. An amorphous powder was obtained and crystallized through slurry in acetone 10 mg/mL at room temperature for three days.

2.3. Single-Crystal X-ray Diffraction (SC-XRD)

Crystals of **TS3w** suitable for SC-XRD analysis were obtained alternatively from slow evaporation of an EtOH solution of **TS3w** and from stratification of diethylether over an EtOH solution of **TS3w**.

Single crystal data were collected at 220 K with a Bruker (US) D8 Venture diffractometer equipped with a Photon II detector, using a CuKα microfocus radiation source (λ = 1.54184 Å). The intensity data were integrated from several series of exposure frames covering the sphere of reciprocal space. Data reductions were performed with APEX3 (III). Absorption corrections were applied using the program SADABS [27]. The structures were solved by intrinsic phasing with the program SHELXT (1.0.825/28 January 2017). Fourier analysis and refinement were performed by the full-matrix least-squares methods based on F^2 using SHELXL-2017 [28,29] implemented in Olex2 software (1.3) [30]. Non-H atoms were refined anisotropically, H atoms were placed in calculated positions and refined with a riding model. Crystallographic data can be found in Supplementary Materials, Table S1.

CCDC 2124344 contains the supplementary crystallographic data for this paper.

2.4. X-ray Powder Diffraction (XRPD)

The crystalline state of samples was investigated by X-ray powder diffraction (XRPD) with an Emyrean Panalytical (UK) V 2.0 instrument equipped with Cu radiation source. Samples were placed on zero background sample holders. The measurements were performed in reflection mode with 2Theta scans from 1.5 to 45°, step size 0.02°, soller slit 0.02 rad, divergence slit 1/8°, antiscatter slit 1/4°. The variable temperature and humidity XRPD analyses were carried out with an Anton Paar (Austria) CHC+ camera equipped with CCU100 temperature control and an MHG-32 humidity generator. The measurements were performed in reflection mode, 2Theta scan from 1.5 to 45°, step size 0.02°, soller slit 0.02 rad, divergence slit 1/8°, antiscatter slit 1/4°.

2.5. Differential Scanning Calorimetry (DSC)

DSC analyses were performed using a routinely calibrated TA Instruments differential scanning calorimeter Discovery equipped with a computer analyzing system (TRIOS). Indium standard and a sapphire disk were used for temperature/enthalpy calibration and heat capacity calibration, respectively. The system was equipped with a refrigerated cooling system (RCS90) accessory under a dry nitrogen purge (50 mL/min). About 1–5 mg of each sample were placed into a Tzero Aluminum hermetic DSC pan covered with a lid. The sample cell was equilibrated at 0 °C and heated under a nitrogen purge (50 mL/min). All samples were given similar thermal histories by linearly heating to 300 °C at a heating rate of 10 °C/min.

2.6. Thermogravimetric Analysis (TGA)

TGA analyses were performed using a TA Instruments Discovery equipped with a computer analyzing system (TRIOS V4.3). About 2 mg of each sample were placed into a Platinum 100 µL pan. The heating rate was 10 °C/min to 300 °C.

2.7. Dynamic Vapour Sorption (DVS)

Moisture sorption/desorption data were collected on a TA Instruments (New Castle, DE, USA) Vapor Sorption Analyzer Q5000SA. First step: sorption data were collected in the range of 40% to 90% relative humidity. Second step: desorption, sorption, and desorption data were collected over a range of 0% to 90% relative humidity (RH) at 10% RH intervals under a nitrogen purge. Samples were not dried prior to the analysis. Equilibrium criteria used for analyses were less than 0.100% weight change in 20 min, with a maximum equilibration time of 1 h if the weight criterion was not met. Data were not corrected for the initial moisture content of the samples. NaBr was used for humidity verification.

2.8. Dissolution Studies

Dissolution studies of theophylline and theophylline squarate trihydrate were carried out using water MilliQ (pH 5.5) at room temperature. 6 suspensions at a concentration of 1.5 mg/mL were prepared for both theophylline and theophylline squarate trihydrate. Each suspension corresponds to a different time point (0–2 min–4 min–6 min–8 min–10 min). From each sample, 200 µL were taken at the corresponding time point and filtered. 20 µL from each taken solution were then diluted (1:10) with H_2O and injected.

Ultra-high-performance liquid chromatography with ultraviolet detection (UHPLC-UV) was used for quantitative analysis of the dissolved drug which absorbs UV-light in the range of 200–290 nm with an absorbance maximum at 254 nm. UHPLC-UV analysis was conducted on a Waters Acquity UPLC system (Milford, MA, USA) that was connected to a diode array detector and equipped with a reversed phase Kinetex® EVO C8 LC-column (100 × 2.1 mm; particle size 1.7 µm; pore size 100 Å, Phenomenex®). The mobile phase consisted of 25 mM of ammonium formiate buffer (pH 3) and 0.1% formic acid in acetonitrile, the flow rate was 0.5 mL/min and the column oven was set to 50 °C. The injection volume was 2 µL.

3. Results

3.1. Theophylline Squarate

3.1.1. Single Crystal Molecular Structure

The asymmetric unit comprises one molecule of theophylline, one molecule of squaric acid, and three molecules of water (See Figure S8). Theophylline exchanges four HBs with two squarate anions via the N11, N21, and C11 atoms (N11···O12, 2.63 Å; N21···O32, 2.64 Å; C11···O22, 3.03 Å; C11···O42, 3.12 Å). Interestingly, C11 forms a bifurcated HB with two squarate anions. In addition, O11 and O21 of Theophylline act as HB acceptors with respect to two symmetry-related O1W water molecules (O21···O1W, 2.77 Å; O11···O1W, 2.92 Å). The two components interact through a HBs connectivity involving two water molecules (O1W and O2W). O1W exchanges three HBs, two as a donor, as already described, and one as acceptor (range 2.74 Å–2.92 Å), O2w exchanges three HBs, two as a donor, and one as acceptor (range 2.53 Å–2.82 Å). O3W forms two HBs as a donor with a symmetry related molecule (3.31 Å) and with O32 of a squarate anion (2.84 Å), respectively, and it accepts HBs from O2W (2.82 Å), Figure 1. On the other hand, O3w is located into a channel like cavity and, according to the long HB distances, it is more loosely bound to the surrounding molecules, Figure S8.

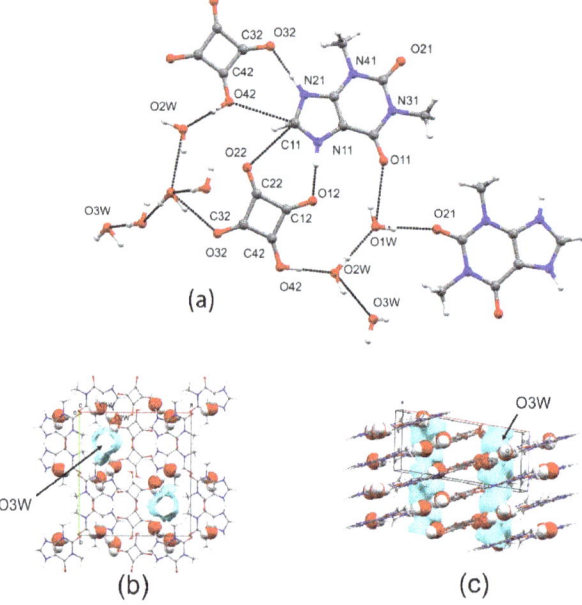

Figure 1. Molecular structure of Theophylline squarate trihydrate with thermal ellipsoids plotted at the 30% probability level, highlighting the intermolecular connection (**a**). The water channel hosting the water molecule O3W is showed along the c-axes (**b**) and along a diagonal direction (**c**).

C–O and C–C bond lengths of the C_4 ring of the squarate are quite indicative of the resonance condition because of the mono-anion formation, Figure 2.

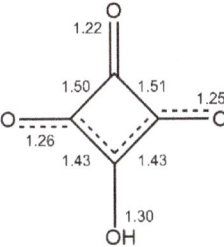

Figure 2. Resonance structure of monodeprotonated squarate anion with bond distance (Å) indicated.

The C42–O42 bond is the longest being the only pure C–O single bond. C22–O22 is the shortest being the only pure double bond. The other two C–O bonds have an intermediate length having a partial double bond character [31]. It follows that C32–C42 and C42–C12 are the shortest covalent bonds associated with a partial double bond.

There are some intrinsic features in the network that the water molecules establish. As previously discussed, O3W is located in a channel that runs parallel to the *c*-axis (see Figure 1b,c). O3W is the most loosely bound to the surroundings and it would be most likely the first molecule leaving the structure upon dehydration. This is also confirmed by the calorimetric analysis as discussed below.

3.1.2. Thermal and Structural Characterization

VT-XRPD (Variable temperature—X-ray powder diffraction) experiment was carried out on **TS3w**. The temperature was slowly increased from to 25 °C to 70 °C–120 °C–155 °C–190 °C as shown in Figure 3. Once water is removed (120 °C) the structure does not collapse and does not lead to the anhydrous form **TSan** either. It dissociates into the starting components. A new diffraction line appears at high temperature (155 °C–190 °C) belonging to a different theophylline polymorph (elusive form V) [32]. However, at the end of the experiment theophylline can be easily detected and no diffraction lines belonging to **TS3w** pattern is observed.

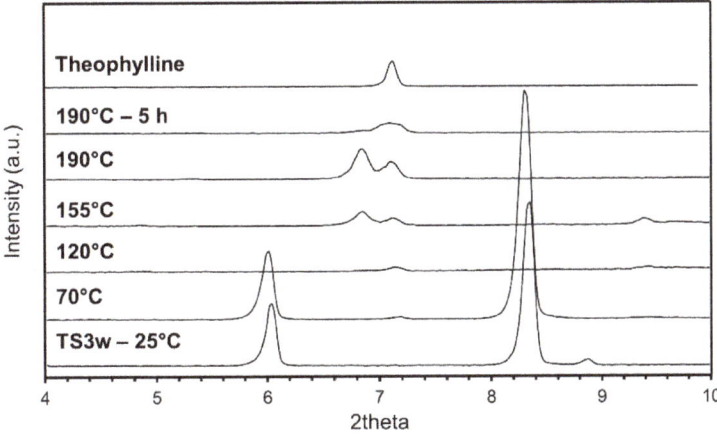

Figure 3. VT-XRPD of Theophylline Squarate trihydrate starting from 25 °C to 190 °C. Above the reference Theophylline.

Thermogravimetric analysis on **TS3w** (See Supplementary Materials, Figure S2) shows a two steps weight loss while increasing the temperature. The first step might represent the channel-water molecule O3W, and the second step might be interpreted as the other two molecules leaving the system, according to the weight changes (5.7% and 10.6%, respectively). DSC thermogram (See Supplementary Materials, Figure S1) shows two endo peaks at 68.9 °C and 89.6 °C, associated with the two events observed in the TGA. VT-XRPD experiment confirms that the structure remains intact at 70 °C. As described above, the salt structure is disrupted only after all of the water molecules have abandoned the HB network (>70 °C), which links Theophylline and Squarate together. The following exothermic event around 138 °C could be interpreted as the rearrangement of the molecules into the two pure components, see Figure 3. **TS3w** melting at 227.7 °C corresponds to the melting point of the anhydrous form (See DSC thermogram of **TSan** in the Supplementary Materials, Figure S5) and it suggests a previous crystallization of **TSan**, potentially happening where the second exothermic peak (170.5 °C) can be detected.

A VH-XRPD (Variable Humidity—X-ray powder diffraction) experiment was also carried out. Theophylline and Squaric Acid (molar ratio 1:1) powders were mixed and exposed to 95% RH for 3 days (Figure 4).

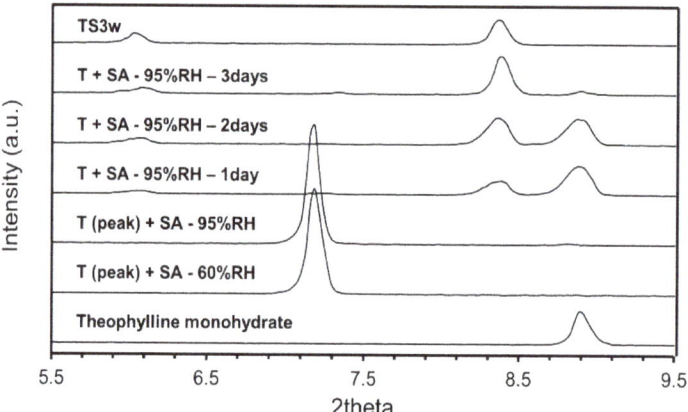

Figure 4. VH-XRPD of Theophylline (T) and Squaric Acid (SA). Below the reference Theo-phylline Monohydrate and above the reference Theophylline squarate trihydrate (**TS3w**).

This experiment confirmed the formation of the salts from the two components in the humidity chamber via a solid-state process without the need of dissolving the components in bulk solvents or applying energy from the milling processes. This observation is complementary to what was observed during the VT-XRPD experiment.

Furthermore, along with the formation of the salts, a steep increase in theophylline monohydrate was observed. After a few days though its signal started to drop while the peak of **TS3w** rose. It can thus be tentatively proposed that the formation of the theophylline monohydrate is a necessary step in the salt formation: as observed in VH-XRPD of pure Theophylline (See Supporting Materials, Figure S9) the monohydrate process starts in less than four hours and is completed in 11 h. The experiment never allowed us to see the complete conversion into the pure theophylline salt. That is probably due to the impossibility of mixing the powder during the experiment being the sample laid on a zero-background XRPD sample stage.

The salt formation has been confirmed also by the FTIR analysis showing the protonated nitrogen signal (See Supplementary Materials, Figure S7).

DVS shows that **TS3w** (See Supplementary Materials, Figure S3) might be considered a stable form since after a cycle of increasing and decreasing humidity, XRPD shows again the same pattern without any loss of crystallinity in the powder (See Supplementary Materials, Figure S4). That proved once again that the anhydrous form could be considered a metastable form.

Figure 5 shows the different powder patterns of commercial Theophylline, Squaric acid, Theophylline squarate trihydrate (**TS3w**), and Theophylline Squarate anhydrous (**TSan**). Thermogravimetric analysis on **TSan** does not show any significant weight loss (See Supplementary Materials, Figures S5 and S6).

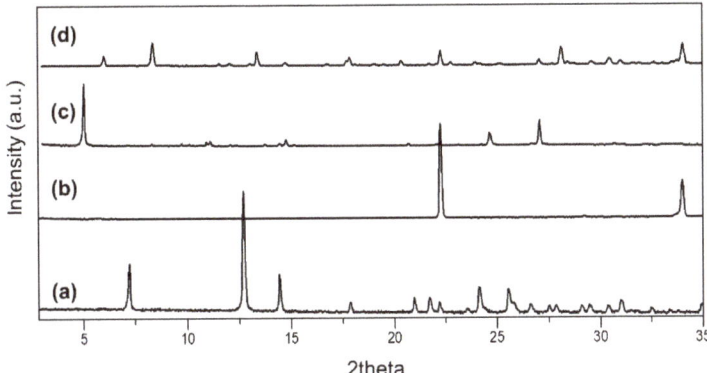

Figure 5. Powder X-ray diffraction of Theophylline (**a**), Squaric acid (**b**), Theophylline Squarate anhydrous (**TSan**, (**c**)) and Theophylline squarate trihydrate (**TS3w**, (**d**)).

3.2. Dissolution Studies

Dissolution studies were carried out to investigate whether the salification would have improved the pharmacokinetic profile in the dissolution step which is mandatory to have absorption of the API. Theophylline alone has an amazingly good bioavailability provided by a very fast and effective dissolution. Nevertheless, we were interested in a slower dissolution profile to better control the release of the drug into the blood stream. This would help to mild the adverse effect when approaching the therapeutic dose.

Those studies demonstrated a better dissolution profile of the salt compared with the free base (Figure 6) leading to a decrease of 52% at 2 min, 54% at 6 min, and 38% at 15 min for **TS3w**. This valuable reduction might turn into an improved in-vivo performance allowing a better control of the administrated drug in terms of both therapeutic effects and adverse reactions. This is very useful in many cases, and it might even be essential when the therapeutic index is very narrow like in this case for Theophylline.

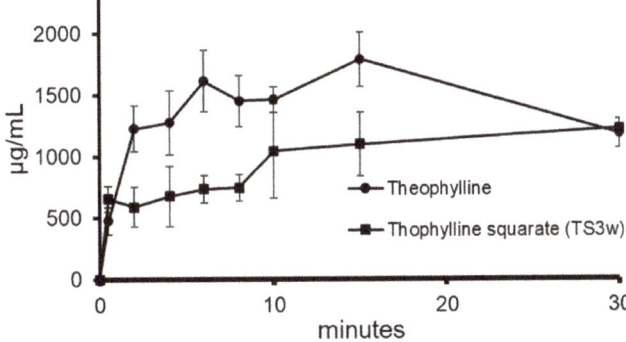

Figure 6. Dissolution study of Theophylline (squares) and Theophylline Squarate trihydrate (rounds).

4. Conclusions

Theophylline salt was treated with squaric acid giving two different forms: anhydrous theophylline squarate and theophylline squarate trihydrate, whose crystal structure has been elucidated allowing to interpret its thermal behavior. Thermal analysis, VH-XRPD, and VT-XRPD experiments allowed us to characterize the stability of **TS3w** and the evolution of the various forms. Indeed, TS3w was stable until 70 °C and it dissociated into the two starting components only above 70 °C, making it suitable for a tablet manufacturing process for instance. Complementary, theophylline, and squaric acid alone led to the salt formation in a high RH% ambient.

During the dissolution study, **TS3w** showed a relevant decrease in the percentage of drugs released in the vehicle. This allows a better control of the administered dose, reducing the adverse effects without losing efficacy.

Salification of Theophylline can thus be considered a valuable strategy to modify its physio-chemical properties and it can be taken into account as an alternative method for controlled release of the drug.

Supplementary Materials: The following are available online at https://www.mdpi.com/article/10.3390/cryst12020201/s1, Table S1. Summary of crystallographic data for Theophylline Squarate trihydrate (TS3w); Figure S1. DSC thermogram of TS3w ; Figure S2. TGA thermogram of TS3w; Figure S3. DVS of TS3w; Figure S4. XRPD of TS3w after DVS (a) compared to its pattern before DVS (b); Figure S5. DSC thermogram of TSan; Figure S6. TGA thermogram of TSan. Figure S7. FTIR of Theophylline (red), Squaric Acid (black) and Theophylline squarate trihydrate (blue). Figure S8. Asymmetric unit of Theophylline squarate trihydrate (a) and molecular structure highlighting the HB connections and distances (b). Figure S9. VH-XRPD of pure Theophylline.

Author Contributions: Project administration, L.M., I.B., F.A. and A.B.; data curation, L.B. and L.F.; formal analysis, L.B. All authors have read and agreed to the published version of the manuscript.

Funding: This research was funded by Chiesi Farmaceutici S.p.A. (CHF-RES-00228).

Institutional Review Board Statement: Not applicable.

Informed Consent Statement: Not applicable.

Conflicts of Interest: The authors declare no conflict of interest.

References

1. Barnes, P.J. Theophylline. *Am. J. Respir. Crit. Care Med.* **2013**, *188*, 901–906. [CrossRef] [PubMed]
2. Focaroli, S.; Jiang, G.; O'connell, P.; Fahy, J.V.; Healy, A.M. The Use of a Three-Fluid Atomising Nozzle in the Production of Spray-Dried Theophylline/Salbutamol Sulphate Powders Intended for Pulmonary Delivery. *Pharmaceutics* **2020**, *12*, 1116. [CrossRef] [PubMed]
3. Barnes, P.J. Theophylline: New Perspectives for an Old Drug. *Am. J. Respir. Crit. Care Med.* **2003**, *167*, 813–818. [CrossRef] [PubMed]
4. Wang, Y.; Wang, C.Z.; Lin, K.X.; Qian, G.S.; Zhuo, W.L.; Li, S.P.; Zhao, Z.Q.; Liao, X.Q.; Song, Y.X. Comparison of Inhaled Corticosteroid Combined with Theophylline and Double-Dose Inhaled Corticosteroid in Moderate to Severe Asthma. *Respirology* **2005**, *10*, 189–195. [CrossRef] [PubMed]
5. Svedmyr, K. Effects of Oral Theophylline Combined with Oral and Inhaled B2-Adrenostimulants in Asthmatics. *Allergy* **1982**, *37*, 119–127. [CrossRef] [PubMed]
6. Pezoa, R.; Gai, M.N.; Gutierrez, C.; Arancibia, A. Development of a Controlled-Release Theophylline Tablet: Evaluation in Vitro and in Vivo. *An. Real Acad. Farm. Inst. Espana* **1992**, *58*, 269–283.
7. Rodrigues, M.; Peiriço, N.; Matos, H.; Gomes De Azevedo, E.; Lobato, M.R.; Almeida, A.J. Microcomposites Theophylline/Hydrogenated Palm Oil from a PGSS Process for Controlled Drug Delivery Systems. *J. Supercrit. Fluids* **2004**, *29*, 175–184. [CrossRef]
8. Jian, H.; Zhu, L.; Zhang, W.; Sun, D.; Jiang, J. Galactomannan (from Gleditsia Sinensis Lam.) and Xanthan Gum Matrix Tablets for Controlled Delivery of Theophylline: In Vitro Drug Release and Swelling Behavior. *Carbohydr. Polym.* **2012**, *87*, 2176–2182. [CrossRef]
9. Buhecha, M.D.; Lansley, A.B.; Somavarapu, S.; Pannala, A.S. Development and Characterization of PLA Nanoparticles for Pulmonary Drug Delivery: Co-Encapsulation of Theophylline and Budesonide, a Hydrophilic and Lipophilic Drug. *J. Drug Deliv. Sci. Technol.* **2019**, *53*, 101128. [CrossRef]

10. Hayashi, T.; Kanbe, H.; Okada, M.; Suzuki, M.; Ikeda, Y.; Onuki, Y.; Kaneko, T.; Sonobe, T. Formulation Study and Drug Release Mechanism of a New Theophylline Sustained-Release Preparation. *Int. J. Pharm.* **2005**, *304*, 91–101. [CrossRef]
11. Vranić, E. Basic Principles of Drug—Excipients Interactions. *Bosn. J. Basic Med. Sci.* **2004**, *4*, 56–58. [CrossRef] [PubMed]
12. West, R. *History of the Oxocarbons*, 1st ed.; Academic Press: New York, NY, USA, 1980; pp. 1–14. [CrossRef]
13. Horiuchi, S.; Tokunaga, Y.; Giovannetti, G.; Picozzi, S.; Itoh, H.; Shimano, R.; Kumai, R.; Tokura, Y. Above-Room-Temperature Ferroelectricity in a Single-Component Molecular Crystal. *Nature* **2010**, *463*, 789–792. [CrossRef] [PubMed]
14. Sreejith, S.; Carol, P.; Chithra, P.; Ajayaghosh, A. Squaraine Dyes: A Mine of Molecular Materials. *J. Mater. Chem.* **2008**, *18*, 264–274. [CrossRef]
15. Ajayaghosh, A. Chemistry of Squaraine-Derived Materials: Near-IR Dyes, Low Band Gap Systems, and Cation Sensors. *Acc. Chem. Res.* **2005**, *38*, 449–459. [CrossRef] [PubMed]
16. Palli, M.A.; McTavish, H.; Kimball, A.; Horn, T.D. Immunotherapy of Recurrent Herpes Labialis with Squaric Acid. *JAMA Dermatol.* **2017**, *153*, 828–829. [CrossRef]
17. Chasák, J.; Šlachtová, V.; Urban, M.; Brulíková, L. Squaric Acid Analogues in Medicinal Chemistry. *Eur. J. Med. Chem.* **2021**, *209*, 112872. [CrossRef] [PubMed]
18. Schwartz, L.M.; Howard, L.O. Aqueous Dissociation of Squaric Acid. *J. Phys. Chem.* **1970**, *74*, 4374–4377. [CrossRef]
19. Karle, I.L.; Ranganathan, D.; Haridas, V. A Persistent Preference for Layer Motifs in Self-Assemblies of Squarates and Hydrogen Squarates by Hydrogen Bonding [X-H···O; X=N, O, or C]: A Crystallographic Study of Five Organic Salts. *J. Am. Chem. Soc.* **1996**, *118*, 7128–7133. [CrossRef]
20. Bertolasi, V.; Gilli, P.; Ferretti, V.; Gilli, G. General Rules for the Packing of Hydrogen-Bonded Crystals as Derived from the Analysis of Squaric Acid Anions: Aminoaromatic Nitrogen Base Co-Crystals. *Acta Crystallogr. Sect. B Struct. Sci.* **2001**, *57*, 591–598. [CrossRef]
21. Jurczak, E.; Mazurek, A.H.; Szeleszczuk, Ł.; Pisklak, D.M.; Zielińska-Pisklak, M. Pharmaceutical Hydrates Analysis—Overview of Methods and Recent Advances. *Pharmaceutics* **2020**, *12*, 959. [CrossRef]
22. Censi, R.; Di Martino, P. Polymorph Impact on the Bioavailability and Stability of Poorly Soluble Drugs. *Molecules* **2015**, *20*, 18759–18776. [CrossRef] [PubMed]
23. Buist, A.R.; Kennedy, A.R.; Manzie, C. Four Salt Phases of Theophylline. *Acta Crystallogr. Sect. C Struct. Chem.* **2014**, *70*, 220–224. [CrossRef] [PubMed]
24. Stevens, J.S.; Byard, S.J.; Schroeder, S.L.M. Salt or Co-Crystal? Determination of Protonation State by X-Ray Photoelectron Spectroscopy (XPS). *J. Pharm. Sci.* **2010**, *99*, 4453–4457. [CrossRef] [PubMed]
25. Mary Novena, L.; Suresh Kumar, S.; Athimoolam, S. Improved Solubility and Bioactivity of Theophylline (a Bronchodilator Drug) through Its New Nitrate Salt Analysed by Experimental and Theoretical Approaches. *J. Mol. Struct.* **2016**, *1116*, 45–55. [CrossRef]
26. Sarma, B.; Saikia, B. Hydrogen Bond Synthon Competition in the Stabilization of Theophylline Cocrystals. *CrystEngComm* **2014**, *16*, 4753–4765. [CrossRef]
27. Sheldrick, G.M. *SADABS-2008/1—Bruker AXS Area Detector Scaling and Absorption Correction*; Bruker AXS: Madison, WI, USA, 2008.
28. Sheldrick, G.M. SHELXT—Integrated Space-Group and Crystal-Structure Determination. *Acta Crystallogr. Sect. A Found. Crystallogr.* **2015**, *71*, 3–8. [CrossRef]
29. Sheldrick, G.M. Crystal Structure Refinement with SHELXL. *Acta Crystallogr. Sect. C Struct. Chem.* **2015**, *71*, 3–8. [CrossRef]
30. Dolomanov, O.V.; Bourhis, L.J.; Gildea, R.J.; Howard, J.A.K.; Puschmann, H. OLEX2: A Complete Structure Solution, Refinement and Analysis Program. *J. Appl. Crystallogr.* **2009**, *42*, 339–341. [CrossRef]
31. Allen, F.H.; Cruz-Cabeza, A.J.; Wood, P.A.; Bardwell, D.A. Hydrogen-Bond Landscapes, Geometry and Energetics of Squaric Acid and Its Mono- and Dianions: A Cambridge Structural Database, IsoStar and Computational Study. *Acta Crystallogr. Sect. B Struct. Sci. Cryst. Eng. Mater.* **2013**, *69*, 514–523. [CrossRef]
32. Fang, C.; Yang, P.; Liu, Y.; Wang, J.; Gao, Z.; Gong, J.; Rohani, S. Ultrasound-Assisted Theophylline Polymorphic Transformation: Selective Polymorph Nucleation, Molecular Mechanism and Kinetics Analysis. *Ultrason. Sonochem.* **2021**, *77*, 105675. [CrossRef]

Article

Lidocaine Pharmaceutical Multicomponent Forms: A Story about the Role of Chloride Ions on Their Stability

Cristóbal Verdugo-Escamilla [1,*], Carolina Alarcón-Payer [2], Francisco Javier Acebedo-Martínez [1], Raquel Fernández-Penas [1], Alicia Domínguez-Martín [3,*] and Duane Choquesillo-Lazarte [1]

1. Laboratorio de Estudios Cristalográficos, IACT-CSIC-Universidad de Granada, Avenida de las Palmeras 4, 18100 Armilla, Spain; j.acebedo@csic.es (F.J.A.-M.); raquel.fernandez@csic.es (R.F.-P.); duane.choquesillo@csic.es (D.C.-L.)
2. Servicio de Farmacia, Hospital Universitario Virgen de las Nieves, 18014 Granada, Spain; carolina.alarconpayer@gmail.com
3. Department of Inorganic Chemistry, Faculty of Pharmacy, University of Granada, 18071 Granada, Spain
* Correspondence: cristobal.verdugo@csic.es (C.V.-E.); adominguez@ugr.es (A.D.-M.)

Abstract: In this investigation, three new crystal forms of lidocaine, and another three of lidocaine hydrochloride with hydroquinone, resorcinol, and pyrogallol were synthesised. All the new forms were characterised using multiple techniques, PXRD, SC-XRD, DSC, and FTIR. The stability of the forms was studied, and, for the more stable forms, i.e., **(lidhcl)** forms, the solubility was determined through FTIR analysis. The new crystalline forms obtained with **(lidhcl)** and the three coformers showed an interesting steric stabilisation mechanism of the oxidation of hydroxybenzenes and showed good physicochemical properties with respect to **(lidhcl)**, constituting a mechanism of modulation of the physicochemical properties.

Keywords: lidocaine; cocrystals; mechanochemistry; multicomponent materials; API; pharmaceutical solids; solubility

1. Introduction

In the past years, crystal engineering and, more particularly, cocrystal design has raised the attention of the pharmaceutical industry as an efficient method to develop new pharmaceutical solid forms. The main advantages of such pharmaceutical cocrystals include not only the enhancement of the physicochemical properties of active pharmaceutical ingredients (APIs) but also the potential of obtaining synergic effects in codrug formulations, keeping the options for intellectual property rights open at lower costs.

Lidocaine (2-diethylamino-N-(2,6-dimethylphenyl)acetamide), hereafter **(lid)**, is an active pharmaceutical ingredient widely used as an anaesthetic in intravenous injection to treat and prevent pain [1,2] in some medical procedures. It is also used in clinics as an antiarrhythmic drug [3] to treat ventricular arrhythmias, specifically ventricular tachycardia and ventricular fibrillation, or as a vasoconstrictor in topical applications [4].

(Lid) shows low solubility in the base form [5]. Therefore, in pharmaceutical formulations, lidocaine is generally used as its hydrochloride derivative **(lidhcl)**. Solubility problems are certainly a big concern regarding the efficacy of oral administration drugs. Hence, if **(lid)** wants to be directly included in drug formulations, one of the best approaches seems to be the development of novel multicomponent pharmaceutical solids, a well-established method able to modulate the physicochemical and biopharmaceutical properties of APIs [6], such as stability, solubility, or manufacturability. In this context, only a few studies can be found in the literature reporting a lidocaine base [7–9], where **(lid)** salts with improved properties were obtained. To build such multicomponent solids, the lidocaine molecule offers different functional groups able to participate in supramolecular synthons (Figure 1), e.g., an amide group that allows hydrogen bonding and an aromatic moiety that can interact through π interactions. These groups are good candidates for interaction with other aromatic rings and alcohol groups, among others.

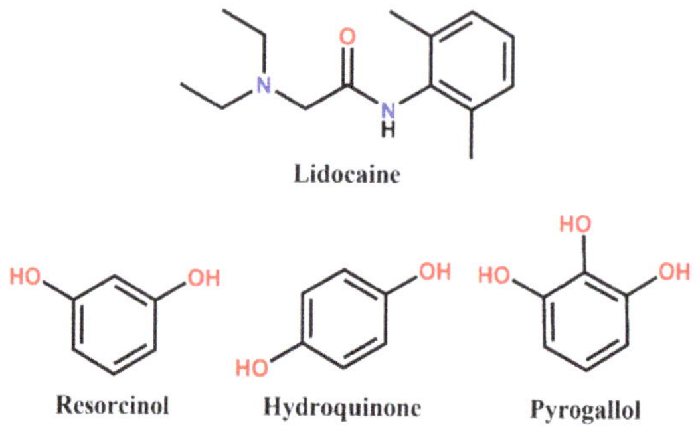

Figure 1. Molecules used in this investigation (lidocaine as base and chlorhydrate).

Polyhydroxy benzenes (Figure 1) are a group of compounds intensely studied and utilised as coformers with multiple APIs including lidocaine [7]. Indeed, they are included in the Generally Recognised as Safe (GRAS) or Substances Added to Food (EAFUS) lists of the US Food and Drug Administration (FDA). Interestingly, these compounds are quite good H-donors; thus, they can easily form hydrogen bonds with other H-acceptor groups such as amide groups, making them excellent candidates to act as coformers in lidocaine formulations, as already reported for phloroglucinol [7], allowing a comparative study of the structural effect of chlorine in the final products [7].

In this work, we focus on comparing the ability of (**lid**) and (**lidhcl**) to cocrystallise with hydroquinone, resorcinol, and pyrogallol, to form multicomponent pharmaceutical solids, and we study how structure can affect physicochemical properties and usability compared to the parent API. The reported multicomponent solids of (**lid**) and (**lidhcl**) were cocrystallised through liquid-assisted grinding (LAG), a versatile and green synthetic method for obtaining solid forms [10,11] that uses mechanical forces to induce chemical transformations, which is a fast and appropriate tool for multicomponent form screening. The resulting multicomponent forms were characterised by powder X-ray diffraction (PXRD), X-ray differential scanning calorimetry/thermogravimetric analysis (DSC/TGA), Fourier-transform infrared spectroscopy (FTIR), and single-crystal X-ray diffraction (SCXRD). In addition, a thorough analysis of the structural details of the corresponding solids forms obtained for (**lid**) and (**lidhcl**) was carried out to unravel the influence of the structure on some relevant physicochemical properties, i.e., solubility and stability.

2. Materials and Methods

All compounds were commercially available from Sigma-Aldrich (St. Louis, MO, USA) and used as received. Solvents were of HPLC grade and were also supplied from Sigma-Aldrich.

2.1. Liquid-Assisted Grinding (LAG)

For the LAG [10,12,13] experiments, different molar ratios of about 100 mg scale and 100 µL of dichloromethane (DCM) were added to each tube. Stoichiometric mixtures of lidocaine (**lid**) and (**lidhc**) with the coformers resorcinol (**res**), hydroquinone (**hq**), and pyrogallol (**pyr**) were gently ground for 30 min at 25 Hz in a Retsch MM400 ball mill (Haan, Germany). A multiple milling homemade accessory allowing for grinding 12 samples at once was used, placing 12 2 mL Eppendorf tubes with three corundum 1 mm balls each.

Three different stoichiometries (1:1, 2:1, 1:2) were screened for each system. The powder materials obtained were analysed by PXRD, FTIR, and DSC/TGA to determine the formation of cocrystals. Cocrystals exhibited distinct PXRD patterns and melting points compared to the starting materials.

2.2. Stability Experiments

In order to investigate the stability with respect to dissociation, a suspension of about 100 mg scale was made with 0.5–1 mL of 0.9% NaCl solution. The suspensions were subjected to magnetic stirring at ambient conditions for 24 h without drying completely, keeping a slurry all the time. Aliquots of the slurry were taken, gently ground, and analysed by PXRD to determine if the cocrystal was dissociated into its components, suffered any transformation, or remained stable in the cocrystal form.

To study the influence of temperature and humidity on the stability of the new phases, these materials were left in a temperature/humidity-controlled chamber with a temperature of 40 °C and 75% relative humidity for 2 months, taking sample aliquots during this time to be analysed by PXRD to evaluate the stability of the crystalline phase. All the samples remained stable as a cocrystal form after 2 months.

2.3. Hetero-Seeding Experiments

In order to obtain single crystals suitable for SCXRD characterisation, evaporation experiments were performed from saturated solutions of the different powders obtained from LAG experiments in DCM. In almost all six cases, single crystals suitable for SCXRD were obtained, except for the **(lid)$_2$(res)** phase, which only formed a microcrystalline material. A hetero-seeding approach was used to obtain single crystals for this phase, using microcrystalline phases obtained for other coformers with predictably similar structures, getting good results using **(lid)$_2$(hq)** as the hetero-seed. This was confirmed after the structure solution because both new cocrystals were isostructural.

For the seeding experiment, a few micrometric solid particles of hetero-seed powder **(lid)$_2$(hq)** were added to the liquified mixture obtained from **(lid)$_2$(res)** LAG experiments, which immediately started to crystallise as single crystals later identified as **(lid)$_2$(res)**.

For the remaining phases, **(lid)$_2$(hq)**, **(lid)$_2$(pyr)**, **(lidhcl)$_2$(hq)**, **(lidhcl)$_2$(res)**, and **(lidhcl)$_2$(pyr)** were obtained by direct recrystallisation from the oily liquid obtained in LAG experiments.

2.4. Powder X-ray Diffraction

PXRD patterns were measured on a Bruker D8 Advance Series II Vario diffractometer (Bruker, AXS, Karlsruhe, Germany) using Cu-K$_{\alpha 1}$ radiation (λ = 1.5406 Å) at 40 kV and 40 mA. Diffraction patterns were collected over 2θ range of 5–60° and using a continuous step size of 0.02° and a total acquisition time of 1 h. The software used for data analysis was Diffrac.EVA v5.0 and TOPAS v6.0 (Bruker, AXS, Karlsruhe, Germany).

2.5. Single-Crystal X-ray Diffraction

Measured crystals were prepared under inert conditions immersed in perfluoropolyether as the protecting oil for manipulation. Suitable crystals were mounted on MiTeGen Micromounts™ (95 Brown Rd, Ithaca, NY, USA) and these samples were used for data collection. Data were collected with a Bruker D8 Venture diffractometer and processed with the APEX3 suite [14]. Structures were solved by direct methods [15], which revealed the position of all non-hydrogen atoms. These atoms were refined on F$_2$ by a full-matrix least-squares procedure using anisotropic displacement parameters [15]. All hydrogen atoms were located by difference Fourier maps and included as fixed contributions riding on attached atoms with isotropic thermal displacement parameters 1.2 times those of the respective atom. Geometric calculations and molecular graphics were performed with Mercury [16] and Olex2 [17]. Additional crystal data are shown in Table 1.

Table 1. Crystallographic information of (lid) and (lidhcl) multicomponent forms.

Compound Name	(lid)$_2$(hq)	(lidhcl)$_2$(hq)	(lid)$_2$(res)	(lidhcl)$_2$(res)	(lid)$_2$(pyr)	(lidhcl)$_2$(pyr)
Formula	C$_{17}$H$_{25}$N$_2$O$_2$	C$_{17}$H$_{26}$ClN$_2$O$_2$	C$_{17}$H$_{25}$N$_2$O$_2$	C$_{17}$H$_{26}$ClN$_2$O$_2$	C$_{26}$H$_{34}$N$_2$O$_7$	C$_{34}$H$_{52}$Cl$_2$N$_4$O$_5$
Formula weight	289.39	325.85	289.39	325.85	486.55	667.69
Crystal system	Triclinic	Monoclinic	Triclinic	Monoclinic	Triclinic	Monoclinic
Space group	P-1	P2$_1$/n	P-1	P2$_1$/n	P-1	P2$_1$/c
a (Å)	a = 7.7056(8)	a = 8.036(2)	a = 7.5096(8)	a = 8.0155(5)	a = 8.5673(5)	a = 10.8407(5)
b (Å)	b = 8.6091(10)	b = 23.252(6)	b = 8.7433(10)	b = 23.2450(16)	b = 12.8458(6)	b = 22.8883(12)
c (Å)	c = 14.1341(17)	c = 10.651(3)	c = 14.2392(15)	c = 10.6788(6)	c = 12.9873(6)	c = 15.7705(8)
α (°)	α = 83.363(5)	α = 90	α = 86.493(7)	α = 90	α = 109.136(2)	α = 90
β (°)	β = 74.552(5)	β = 111.791(14)	β = 75.831(6)	β = 111.785(3)	β = 105.849(2)	β = 108.500(2)
γ (°)	γ = 69.625(4)	γ = 90	γ = 72.311(6)	γ = 90	γ = 100.344(2)	γ = 90
V (Å^3)	846.94(17)	1848.0(9)	863.54(17)	1847.6(2)	1240.75(11)	3710.8(3)
Z	2	4	2	4	2	4
D$_c$ (g·cm^{-3})	1.135	1.171	1.113	1.113	1.302	1.195
μ (mm^{-1})	0.590	1.892	0.579	1.892	0.779	1.917
F(000)	314	700	314	700	520	1432
Reflections collected	11839	14214	14062	14524	17040	41420
Unique reflections	2951	3236	2995	3221	4245	6519
R$_{int}$	0.1392	0.0635	0.0254	0.0531	0.0387	0.0742
Data/restraints/parameters	2951/0/196	3236/0/204	2995/41/314	3221/53/242	4245/0/326	6519/0/418
Goodness of fit (F^2)	1.070	1.081	1.094	1.064	1.035	1.060
R$_1$ (I > 2σ(I))	0.0688	0.0487	0.0645	0.0457	0.0397	0.0524
wR$_2$ (I > 2σ(I))	0.2007	0.1424	0.2053	0.1202	0.1024	0.1389
CCDC number	2125120	2125121	2125122	2125123	2125124	2125125

2.6. Thermal Analysis

For the DSC/TGA experiments, samples in the range of 30 mg were studied using a Mettler Toledo TGA/DSC 3+ Star analyser. Samples were heated at 10 °C/min in the temperature range 25–190 °C under a nitrogen atmosphere with 100 mL/min flow in aluminium capsules.

2.7. Fourier-Transform Infrared Spectra

Fourier-transform infrared spectra (FTIR) were recorded with an attenuated total reflectance (ATR) accessory diamond crystal using an Invenio R FTIR spectrometer (Bruker). FTIR spectra were recorded within the wavenumber range from 4000 cm^{-1} to 400 cm^{-1} at 2 cm^{-1} resolution. In order to correctly subtract the background and, hence, obtain less noisy spectra, the solvent (0.9% NaCl water solution) was used at room temperature for the background measurement.

2.8. Solubility Assays

Thermodynamic solubility measurements were performed in an Invenio R FTIR spectrometer (Bruker) after equilibrating the solids in a 0.9% NaCl water solution [8] under stirring at 500 rpm for 24 h. After the equilibrating time, the suspensions were filtrated through a 20 μm filter, and the resulting clear solution was analysed by FTIR. Then, the solid was analysed by PXRD to study the phase stability after equilibrium, resulting for (lidhcl)$_2$(res) and (lidhcl)$_2$(hq) that the cocrystal form remained stable after 24 h but not for (lidhcl)$_2$(pyr). A calibration curve was built with different (lidhcl) concentrations [18,19] obtaining linear models with R^2 values greater than 0.99, which were used to calculate the thermodynamic solubility. The peak used for the calibration curves was the area between 1712 cm^{-1} and 1612 cm^{-1}, corresponding to (lidhcl), where there was no interference of any coformer.

3. Results and Discussion

Six new phases were obtained for (lid) and (lidhcl) with hydroquinone, resorcinol, and pyrogallol. Only the phases with (lidhcl) demonstrated improved stability; accordingly, the physical characterisation is focused on these three new phases, after which the six phases are characterised structurally in order to study this stability differences. In all the cases, the correspondence between the SCXRD solved structures and the bulk powder was confirmed by PXRD (see Supplementary Figure S1).

3.1. Stability Studies

After preparing slurries of the three new (lidhcl) phases, it can be clearly seen (Figure S2) that (lidhcl)$_2$(res) and (lidhcl)$_2$(hq) were stable after 24 h stirring in water but not (lidhcl)$_2$(pyr), which transformed into a new phase that could not be structurally characterised but only identified as new form. Moreover, it was observed that crystal forms with (lid), compared to (lidhcl), drastically changed their colour (Figure S2), which is a clear indicative of their poor stability with respect to oxidation.

Supplementary experiments were performed to evaluate the stability of (lidhcl) forms, in ageing conditions (40 °C and 75% relative humidity), where it can be observed that all three (lidhcl) new forms were stable after 2 months in ageing conditions (Figure 2), which did not occur for (lid) forms, as can be easily seen from Figure S2.

3.2. FTIR Spectroscopy and Solubility Measurements

Figure 3 shows the FTIR spectra for all six new phases obtained in this investigation. FTIR provides a fingerprint sign of each compound and wealthy information about the noncovalent interactions between acceptor and donor groups. Peak shifts can be found in the bands of the functional groups involved in the hydrogen bonds, namely, the carbonyl –C=O functional group of the amine group of the lidocaine and lidocaine hydrochloride molecule and –OH groups in the polyphenol molecules. The shifts in lidocaine's –C=O group stretching vibration occurred from 1661 cm^{-1} to 1623–1620 cm^{-1} for the new crystalline forms obtained, and from 1680 cm^{-1} to 1677–1670 cm^{-1} for lidocaine hydrochloride. These shifts demonstrate the hydrogen bonding between the –C=O group of lidocaine molecules and the –OH groups of the polyphenols [7], which as further confirmed by the structure solution.

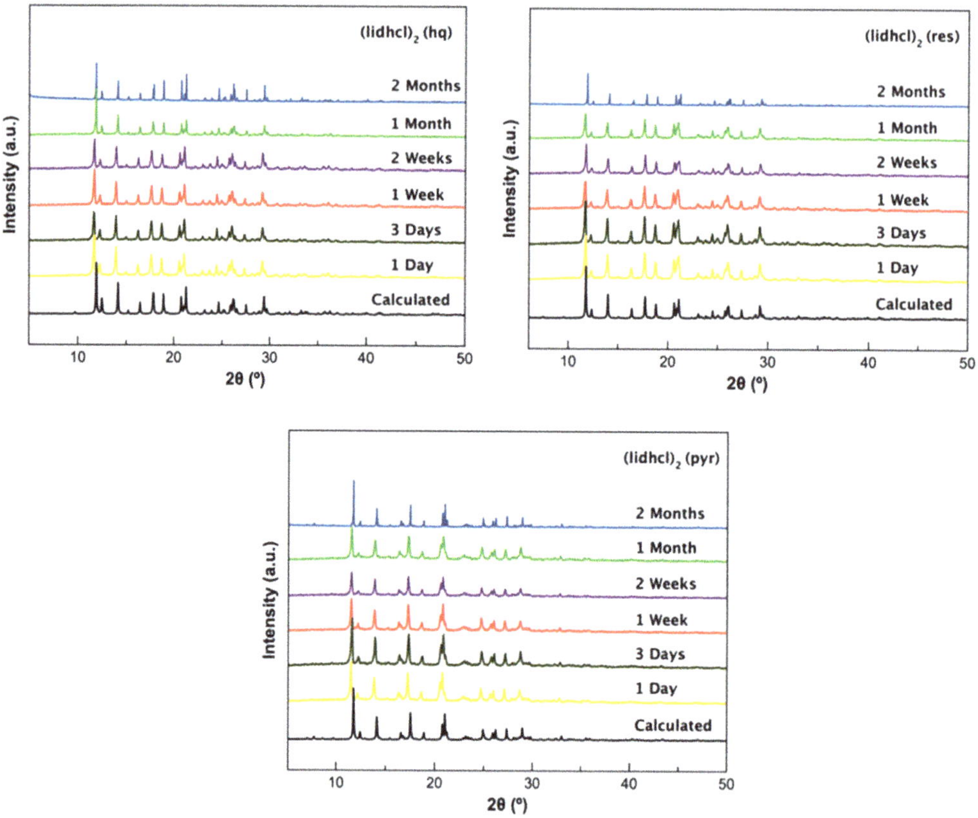

Figure 2. Stability in ageing conditions (40 °C and 75% relative humidity) for the three **(lidhcl)** new forms.

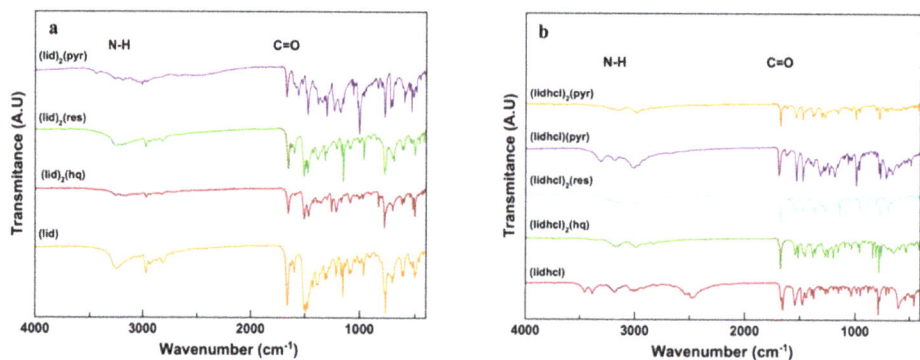

Figure 3. FTIR spectra of (**a**) (**lid**) and (**b**) (**lidhcl**) multicomponent forms.

In order to measure the solubility of the new (lidhcl) forms, FTIR measurements were performed to avoid the overlapping observed in the UV spectra, due to the ability of FTIR to show us isolated peaks from each component, allowing us to construct a calibration curve with the intensity of an unambiguous peak of (lidhcl).

A calibration curve was built with different (lidhcl) concentrations [18,19] obtaining linear models with R^2 values greater than 0.99 (Figure 4), which were used to calculate the apparent solubility. The peak used for the calibration curves was the area between 1712 cm^{-1} and 1612 cm^{-1} corresponding to (lidhcl), where there was no interference of any of the used coformers.

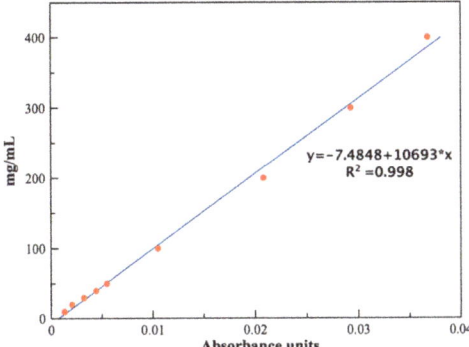

Figure 4. Calibration curve of (**lidhcl**) concentrations ranging from 20 to 400 mg/mL.

The constructed calibration curve is shown in Figure 4, where the model's excellent agreement with the experimental concentrations can be appreciated. Each concentration was prepared and measured in duplicate to assure reproducibility, and the mean value is represented.

Table 2 presents the apparent solubilities calculated from FTIR data for each new (lidhcl) phase.

Table 2. . Solubilities in (mg/mL) of (**lidhcl**) multicomponent forms obtained from FTIR data.

(lidhcl)2(hq)	244.7
(lidhcl)2(res)	241.0
(lidhcl)2(pyr)	126.0

The solubility decreased in the new phases with respect to reported values for (**lidhcl**) [9], generating (**lidhcl**) forms with an interesting path to modulate solubility.

3.3. X-ray Diffraction

Powder X-ray diffraction patterns are a distinctive fingerprint of the crystalline phases. It can be confirmed from Figure 5 how the PXRD patterns of the new (**lid**) and (**lidhcl**) phases were entirely different from the starting API, ensuring the appearance of new crystalline phases. Additionally, the comparison of the simulated PXRD data from SCXRD solved structures with the observed patterns obtained for each bulk powder obtained from LAG experiments could confirm the new obtained phases (Figure S1).

From the single crystals obtained for each new phase, structure solutions were achieved, and these structures were analysed to extract information that could relate stability to the structure.

Cocrystals of lidocaine free base (**lid**) with two di-hydroxy isomers (hydroquinone, **hq** and resorcinol, **res**) and one tri-hydroxybenzene (pyrogallol, **pyr**) coformer were cocrystallised, and their crystal structure was determined. To evaluate the effect of the chloride ion on the stability and properties performance of lidocaine multicomponent solid forms, analogue cocrystals were also obtained using lidocaine hydrochloride salt **(lidhcl)**. The structural features of both free base and ionic lidocaine systems are discussed in this section.

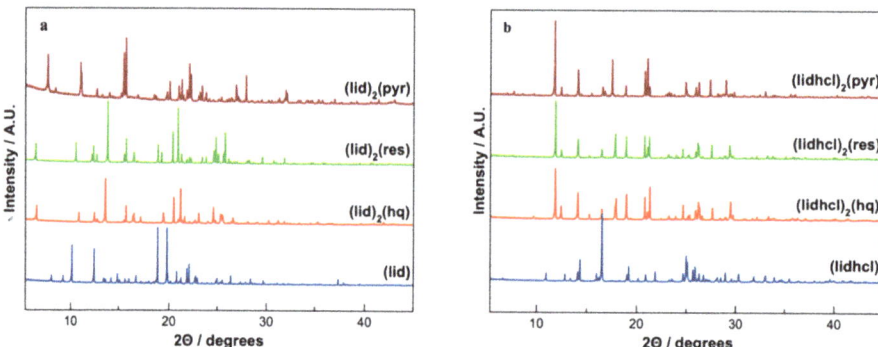

Figure 5. PXRD patterns of (**a**) **(lid)** and (**b**) **(lidhcl)** new multicomponent forms.

3.3.1. (lid/hq) Systems

The asymmetric unit of **(lid)₂(hq)** contains one **(lid)** molecule and half an **(hq)** molecule located at an inversion centre, leading to a 2:1 stoichiometric ratio (Figure 6a). The hydroxy groups of **(hq)** point in opposite directions, thus adopting a *trans* conformation, forming discrete intermolecular OH(phenol)···O(carbonyl) hydrogen bonds with the $D_1^1(2)$ graph set between the OH donor **(hq)** and the carbonyl oxygen acceptor of the **(lid)**. These discrete units are further connected by centrosymmetric –C–H···π (2.968 Å) and –N–H···π (3.254 Å) interactions between **(lid)** molecules, generating a 1D chain (Figure 6b). Hydrophobic interactions involving methyl groups associate chains to build up the 3D structure.

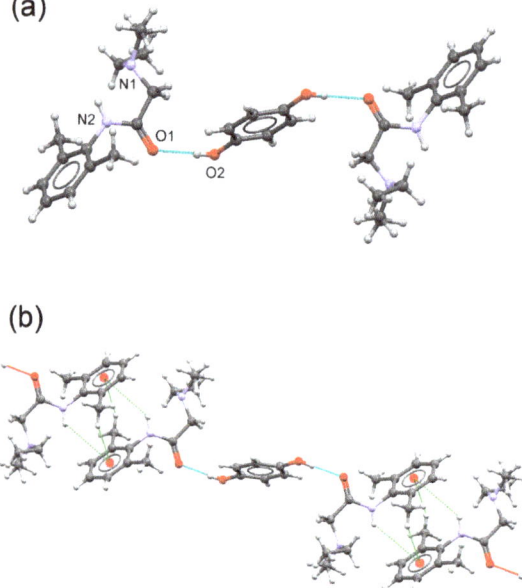

Figure 6. (**a**) Asymmetric unit of **(lid)₂(hq)** cocrystal. (**b**) Detailed view of the –C–H···π and –N–H···π interactions that connect discrete **(lid)₂(hq)** units to form a chain structure.

The asymmetric unit of **(lidhcl)₂(hq)** contains one **(lidhcl)** salt and half an **(hq)** molecule located at an inversion centre, with a 2:2:1 stoichiometric ratio considering all the components (Figure 7a). Protonation of the N1 amine group results in a change in the **(lid)** conformation, locating the amine groups in *trans* conformation. The N2 amine group and the ammonium N1 group participate in electrostatic hydrogen bonds with the chloride ion, generating a chain reinforced by C–H···O hydrogen bonds. The **(hq)** molecules connect chains via additional hydrogen bonds involving chloride ions, generating a 3D structure. The participation of chloride ions through noncovalent interactions results in a more compact crystal structure than the free base cocrystal analogue where each **(hq)** molecule is surrounded by six **(lid⁺)** cations, protecting the OH groups of **(hq)** from oxidation (Figure 7b).

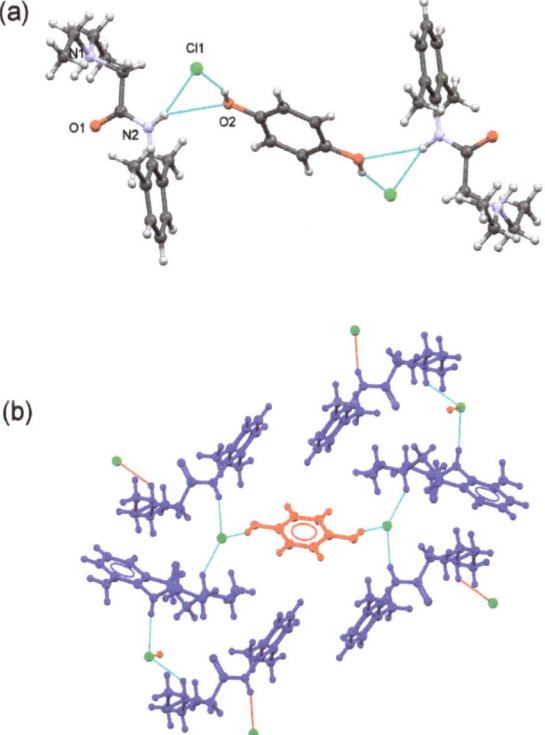

Figure 7. (a) Asymmetric unit of **(lidhcl)₂(hq)** ionic cocrystal. (b) Detailed view of the **(hq)** environment in the crystal structure. Blue: **(lid⁺)** ion, green: Cl⁻ ion, red: **(hq)** molecule.

3.3.2. (lid/res) Systems

(lid)₂(res) crystallises in the triclinic P-1 space group. As in the **(lid)₂(hq)** cocrystal, the coformer is sitting on an inversion centre; however, in this compound, the **(res)** coformer molecular symmetry does not exhibit inversion. Therefore, a disorder of the molecule about this particular special position is observed. The **(res)** adopts a syn–syn conformation, pointing the –OH groups through the –C=O group of **(lid)** ($D_1^1(2)$ graph set) and establishing a 2:1 stoichiometric ratio (Figure 8). These discrete units are connected in a similar way than in **(lid)₂(hq)** through shorter –C–H···π (2.864 Å) and –N–H···π (3.216 Å) interactions between **(lid)** molecules, generating chains. Again, hydrophobic interactions involving methyl groups associate chains to build up the 3D structure. Crystal structure similarities in the **(lid)** cocrystals [20] with **(hq)** and **(res)** coformers could explain the template effect of **(lid)₂(hq)** solid when used as hetero-seeds during the synthesis of crystals of **(lid)₂(res)**

(see Section 2.3). The results obtained from crystal packing similarity calculations of one of the alternative positions of the disordered **(lid)$_2$(res)** structure and **(lid)$_2$(hq)** structure using Mercury [21] showed that 14 out of 20 molecules were matched in the pairs of 2:1 cocrystals with a PXRD similarity index of 0.96745 and RMSD of 0.268 (Figure 9). These results suggest that these pairs possess identical intermolecular interactions and lead to the same crystal packing [22].

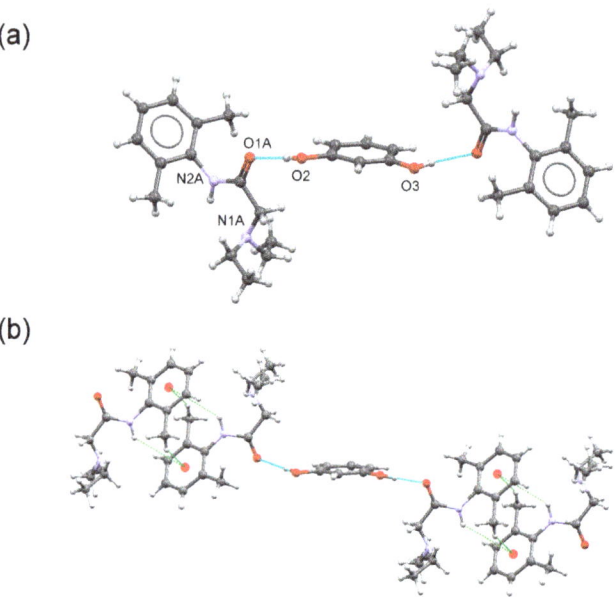

Figure 8. (**a**) Asymmetric unit of **(lid)$_2$(res)** cocrystal. Only one of the two alternative positions in the disordered structure is shown for clarity. (**b**) Detailed view of the –C–H···π and –N–H···π interactions that connect discrete **(lid)(res)** units to form a chain structure.

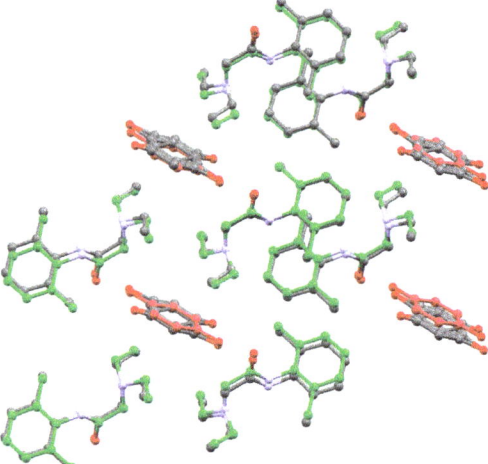

Figure 9. Crystal packing similarity plot comparing the **(lid)$_2$(hq)** and **(lid)$_2$(res)** crystal structures. Green and red: **lid-hq** molecules.

The **(lidhcl)₂(res)** ionic cocrystal [20] contains one **(lidhcl)** salt and half a **(res)** molecule located in an inversion centre, giving a 2:1:1 stoichiometric ratio (Figure 10a). As observed in the ionic cocrystal of **(lidhcl)** with **(hq)**, protonation of **(lid)** imposes a conformational change resulting in a chain structure build-up by electrostatic hydrogen bonds involving protonated lidocaine donor and acceptor groups, as well as chloride ions. Disordered **(res)** molecules connect chains to generate the supramolecular 3D crystal structure. Similarly, as in the ionic cocrystal **(lidhcl)₂(hq)**, **(res)** coformer molecules are protected with **(lid⁺)** cations, preventing them from oxidation (Figure 10b).

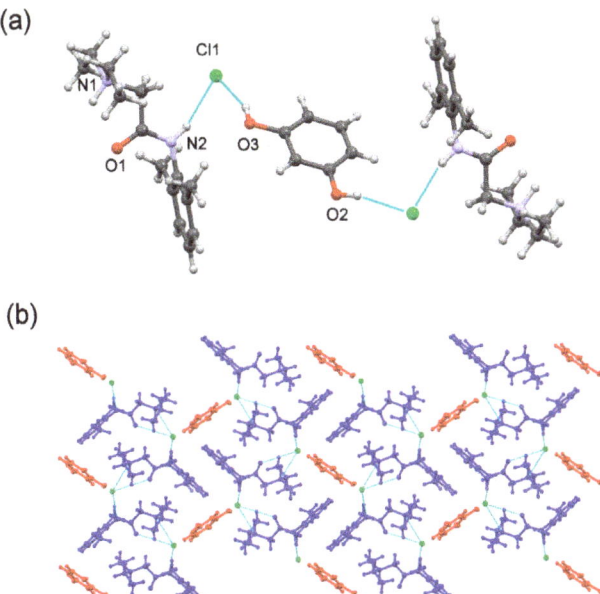

Figure 10. (a) Asymmetric unit of **(lidhcl)₂(res)** ionic cocrystal. Only one of the two alternative positions in the disordered structure is shown for clarity. (b) Detailed view of the crystal packing of **(lidhcl)₂(res)** ionic cocrystal (view along the a axis). Blue: **(lid⁺)** ion, green: Cl⁻ ion, red: **(res)** molecules.

3.3.3. (lid/pyr) Systems

The **(lid)₂(pyr)** is an ionic cocrystal and crystallises in the triclinic P-1 space group. Its asymmetric unit contains one **(lid⁺)** molecule, one **(pyr-)** anion, and one **(pyr)** neutral molecule in a stoichiometric ratio 1:1:1 (Figure 11a). Protonation of the tertiary amine in **(lid)** imposes a change in its molecular conformation, pointing both amine protons in opposite directions with –N–C–C–N– torsion angles of −157.97° for **(lid)₂(pyr)** in comparison with −11.27° and 6.41° for **(lid)₂(hq)** and **(lid)₂(res)**, respectively. The **(pyr)** molecules adopt the anti-conformation [22]. This conformation is energetically the least stable. However, considering the packing arrangement in **(lid)₂(pyr)**, the anti-conformation ensures that each hydroxyl group of the pyrogallol moiety participates in hydrogen bonding interactions. In the crystal, **(pyr)** molecules are connected to three **(pyr-)** anions through H-bonding interactions generating chains running along the a axis. The **(lid⁺)** molecules intercalate between two adjacent **(pyr⁻)** anions through N–H⋯O(hydroxyl) and O–H⋯O(carbonyl) H-bonding and C–H interactions (Figure 11b). In the crystal, C–H⋯π interactions connect these supramolecular chains to form the 3D structure.

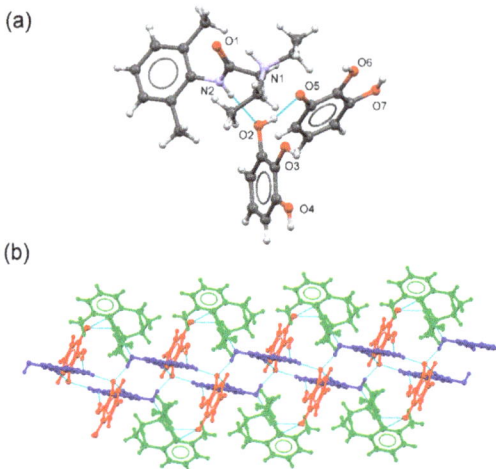

Figure 11. (a) Asymmetric unit of **(lid)₂(pyr)** ionic cocrystal. (b) Fragment of the supramolecular chain running along the a axis. Green: **(lid⁺)** cations, red: **(pyr⁻)** anions, blue: **(pyr)** molecules.

The **(lidhcl)₂(pyr)** is also an ionic cocrystal and crystallises in the monoclinic P21/c space group. Its asymmetric unit contains two **(lid⁺)** two chloride ions and one **(pyr)** molecule in a stoichiometric ratio 2:2:1 (Figure 12a). As observed in the free base ionic cocrystal, **(pyr)** adopts an anti-conformation where each hydroxyl group forms a hydrogen bond interaction with the corresponding chloride ion. In the crystal, Cl1A and Cl1B ions are H-bonded to four molecules, two **(pyr)** and two **(lid⁺)**, and three molecules, one **(pyr)** and two **(lid⁺)** ions, respectively. This arrangement leads to efficient packing, shielding pairs of **(pyr)** molecules that are surrounded with **(lid⁺)** and Cl⁻ ions, thus protecting the coformer from oxidation (Figure 12b).

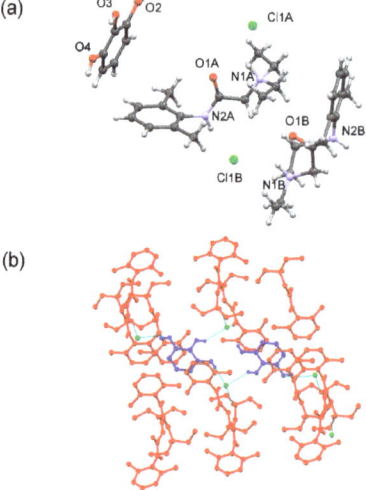

Figure 12. (a) Asymmetric unit of **(lidhcl)₂(pyr)** ionic cocrystal. (b) Detailed view of the environment of a pair of **(pyr)** molecules in the crystal structure. Red: **(lid⁺)** cations, blue: **(pyr)** molecules, green: Cl⁻ ions.

3.4. Thermal Analysis

DSC experiments were performed to evaluate the stability of the new phases, as well as to determine the melting points of the new multicomponent phases. Figure 13 shows the DSC of the three **(lidhcl)** and **(lid)** phases.

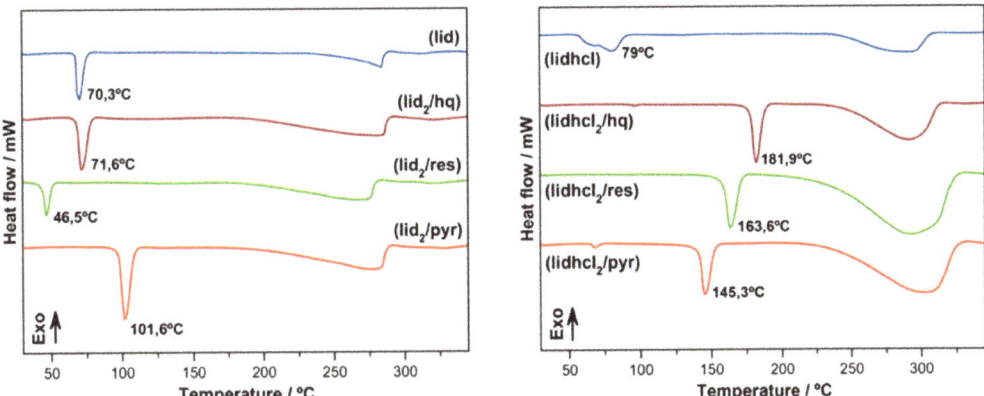

Figure 13. DSC diagrams for **(lid)** phases (**left**) and for **(lidhcl)** phases (**right**).

In the case of the **(lidhcl)** phases, a decrease in the melting point can clearly be seen in the order **(hq)** > **(res)** > **(pyr)**, with apparently no relation with structure effects.

There was a certain tendency with the measured solubilities, which also decreased in the same order, relating the melting point with the apparent solubility.

4. Conclusions

In this investigation, six new multicomponent forms, three for lidocaine and three for lidocaine hydrochloride, with hydroquinone, resorcinol, and pyrogallol, were obtained and characterised by different techniques. LAG allowed the synthesis of six new crystal forms, and the stability assays revealed the high stability of the **(lidhcl)** forms compared with **(lid)** phases, which were selected for further analysis. The SC-XRD structure solution allowed a comparative structural analysis, highlighting how **(lidhcl)** increases the stability of the new forms compared to **(lid)** through a steric protection effect. The structural information obtained revealed the important role that chloride ions play in the stabilisation of the **(lidhcl)** new multicomponent forms, allowing an improved oxidation behaviour.

There was a tendency between melting points and solubility for the new **(lidhcl)** phases. The solubility of the new crystal forms of **(lidhcl)** was lower than **(lidhcl)** itself, which, far from being a drawback, can be an interesting way to modulate the **(lidhcl)** solubility and stability, as can be concluded from the DSC of the new phases showing an increase in thermal stability. In this concern, the synthesis of molecular or ionic cocrystals can be taken as a valuable approach to tuning the properties of the API to adapt it to the requirements.

Supplementary Materials: The following are available online at https://www.mdpi.com/article/10.3390/cryst12060798/s1: Figure S1. Calculated vs. experimental PXRD patterns for the six new phases; Figure S2. **(lidhcl)** vs. **(lid)** new phases: visual comparative slurry stabilities. Left: **(lidhcl)** forms, right: **(lid)** forms.

Author Contributions: Conceptualisation and methodology, C.V.-E. and D.C.-L.; formal analysis and investigation, C.V.-E., C.A.-P., F.J.A.-M., R.F.-P., D.C.-L. and A.D.-M.; writing—original draft preparation, C.V.-E.; writing—review and editing, C.V.-E. and A.D.-M.; funding acquisition, D.C.-L. and A.D.-M.; supervision, C.V.-E. and A.D.-M. All authors have read and agreed to the published version of the manuscript.

Funding: This research was funded by the Spanish Agencia Estatal de Investigación of the Ministerio de Ciencia, Innovación, y Universidades (MICIU) and co-funded by FEDER, UE, Project PGC2018-102047-B-I00 (MCIU/AEI/FEDER, UE) and Project B-FQM-478-UGR20 (FEDER-Universidad de Granada, Spain).

Institutional Review Board Statement: Not applicable.

Informed Consent Statement: Not applicable.

Data Availability Statement: Not applicable.

Acknowledgments: F.J.A.-M. wants to acknowledge an FPI grant (Ref. PRE2019-088832). C.V.-E. acknowledges Project PTA2020-019483-I funded by the Spanish Agencia Estatal de Investigación of the Ministerio de Ciencia e Innovación.

Conflicts of Interest: The authors declare no conflict of interest.

References

1. Foo, I.; Macfarlane, A.J.R.; Srivastava, D.; Bhaskar, A.; Barker, H.; Knaggs, R.; Eipe, N.; Smith, A.F. The Use of Intravenous Lidocaine for Postoperative Pain and Recovery: International Consensus Statement on Efficacy and Safety. *Anaesthesia* **2021**, *76*, 238–250. [CrossRef]
2. Bigger, J.T., Jr.; Giardina, E.-G.V. The Pharmacology and Clinical Use of Lidocaine and Procainamide. *MCV Q.* **1973**, *9*, 65–76.
3. Collinsworth, K.A.; Kalman, S.M.; Harrison, D.C. The Clinical Pharmacology of Lidocaine as an Antiarrhythmic Drug. *Circulation* **1974**, *50*, 1217–1230. [CrossRef] [PubMed]
4. YuvalBloch, M.D. Use of Topical Application of Lidocaine-Prilocaine Cream to Reduce Injection-Site Pain of Depot Antipsychotics. *Psychiatr. Serv.* **2004**, *55*, 940–941. [CrossRef]
5. Badawi, H.M.; Förner, W.; Ali, S.A. The Molecular Structure and Vibrational, 1H and 13C NMR Spectra of Lidocaine Hydrochloride Monohydrate. *Spectrochim. Acta Part A Mol. Biomol. Spectrosc.* **2016**, *152*, 92–100. [CrossRef] [PubMed]
6. Sathisaran, I.; Dalvi, S.V. Engineering Cocrystals of Poorly water-Soluble Drugs to Enhance Dissolution in Aqueous Medium. *Pharmaceutics* **2018**, *10*, 108. [CrossRef] [PubMed]
7. Magaña-Vergara, N.E.; de La Cruz-Cruz, P.; Peraza-Campos, A.L.; Martínez-Martínez, F.J.; González-González, J.S. Mechanochemical Synthesis and Crystal Structure of the Lidocaine-Phloroglucinol Hydrate 1:1:1 Complex. *Crystals* **2018**, *8*, 130. [CrossRef]
8. Braga, D.; Chelazzi, L.; Grepioni, F.; Dichiarante, E.; Chierotti, M.R.; Gobetto, R. Molecular Salts of Anesthetic Lidocaine with Dicarboxylic Acids: Solid-State Properties and a Combined Structural and Spectroscopic Study. *Cryst. Growth Des.* **2013**, *13*, 2564–2572. [CrossRef]
9. Groningsson, K.; Lindgren, J.-E.; Lundberg, E.; Sandberg, R. Lidocaine Base and Hydrochloride. In *Analytical Profiles of Drug Substances*; Academic Press: Washington, DC, USA, 1985.
10. Ying, P.; Yu, J.; Su, W. Liquid-Assisted Grinding Mechanochemistry in the Synthesis of Pharmaceuticals. *Adv. Synth. Catal.* **2021**, *363*, 1246–1271. [CrossRef]
11. Howard, J.L.; Sagatov, Y.; Repusseau, L.; Schotten, C.; Browne, D.L. Controlling Reactivity through Liquid Assisted Grinding: The Curious Case of Mechanochemical Fluorination. *Green Chem.* **2017**, *19*, 2798–2802. [CrossRef]
12. Do, J.L.; Friščić, T. Mechanochemistry: A Force of Synthesis. *ACS Cent. Sci.* **2017**, *3*, 13–19. [CrossRef]
13. Bruker APEX3. *APEX3 V2019.1*; Bruker AXS Inc.: Madison, WI, USA, 2019.
14. Sheldrick, G.M. SHELXT—Integrated Space-Group and Crystal-Structure Determination. *Acta Crystallogr. Sect. A Found. Crystallogr.* **2015**, *71*, 3–8. [CrossRef] [PubMed]
15. Macrae, C.F.; Bruno, I.J.; Chisholm, J.A.; Edgington, P.R.; McCabe, P.; Pidcock, E.; Rodriguez-Monge, L.; Taylor, R.; van de Streek, J.; Wood, P.A. Mercury CSD 2.0—New Features for the Visualization and Investigation of Crystal Structures. *J. Appl. Crystallogr.* **2008**, *41*, 466–470. [CrossRef]
16. Dolomanov, O.V.; Bourhis, L.J.; Gildea, R.J.; Howard, J.A.K.; Puschmann, H. OLEX2: A Complete Structure Solution, Refinement and Analysis Program. *J. Appl. Crystallogr.* **2009**, *42*, 339–341. [CrossRef]
17. Togkalidou, T.; Fujiwara, M.; Patel, S.; Braatz, R.D. Solute Concentration Prediction Using Chemometrics and ATR-FTIR Spectroscopy. *J. Cryst. Growth* **2001**, *231*, 534–543. [CrossRef]
18. Hojjati, H.; Rohani, S. Measurement and Prediction of Solubility of Paracetamol in Water-Isopropanol Solution. Part 1. Measurement and Data Analysis. *Org. Process Res. Dev.* **2006**, *10*, 1101–1109. [CrossRef]
19. Aitipamula, S.; Banerjee, R.; Bansal, A.K.; Biradha, K.; Cheney, M.L.; Choudhury, A.R.; Desiraju, G.R.; Dikundwar, A.G.; Dubey, R.; Duggirala, N.; et al. Polymorphs, Salts, and Cocrystals: What's in a Name? *Cryst. Growth Des.* **2012**, *12*, 2147–2152. [CrossRef]
20. MacRae, C.F.; Sovago, I.; Cottrell, S.J.; Galek, P.T.A.; McCabe, P.; Pidcock, E.; Platings, M.; Shields, G.P.; Stevens, J.S.; Towler, M.; et al. Mercury 4.0: From Visualization to Analysis, Design and Prediction. *J. Appl. Crystallogr.* **2020**, *53*, 226–235. [CrossRef] [PubMed]

21. Ranjan, S.; Devarapalli, R.; Kundu, S.; Saha, S.; Deolka, S.; Vangala, V.R.; Reddy, C.M. Isomorphism: "Molecular Similarity to Crystal Structure Similarity" in Multicomponent Forms of Analgesic Drugs Tolfenamic and Mefenamic Acid. *IUCrJ* **2020**, *7*, 173–183. [CrossRef] [PubMed]
22. Vedernikova, I.; Salahub, D.; Proynov, E. DFT Study of Hyperconjugation Effects on the Charge Distribution in Pyrogallol. *J. Mol. Struct. THEOCHEM* **2003**, *663*, 59–71. [CrossRef]

Article

Towards the Development of Novel Diclofenac Multicomponent Pharmaceutical Solids

Francisco Javier Acebedo-Martínez [1], Carolina Alarcón-Payer [2], Helena María Barrales-Ruiz [1,3], Juan Niclós-Gutiérrez [3], Alicia Domínguez-Martín [3] and Duane Choquesillo-Lazarte [1,*]

[1] Laboratorio de Estudios Cristalográficos, IACT, CSIC-Universidad de Granada, Avda. de las Palmeras 4, 18100 Armilla, Spain; j.acebedo@csic.es (F.J.A.-M.); helenabarrales@gmail.com (H.M.B.-R.)
[2] Servicio de Farmacia, Hospital Universitario Virgen de las Nieves, 18014 Granada, Spain; carolina.alarconpayer@gmail.com
[3] Departament of Inorganic Chemistry, Faculty of Pharmacy, University of Granada, 18071 Granada, Spain; jniclos@ugr.es (J.N.-G.); adominguez@ugr.es (A.D.-M.)
* Correspondence: duane.choquesillo@csic.es

Abstract: Multicomponent pharmaceutical materials offer new opportunities to address drug physicochemical issues and to obtain improved drug formulation, especially on oral administration drugs. This work reports three new multicomponent pharmaceutical crystals of the non-steroidal anti-inflammatory drug diclofenac and the nucleobases adenine, cytosine, and isocytosine. They have been synthesized by mechanochemical methods and been characterized in-depth in solid-state by powder and single crystal X-ray diffraction, as well as other techniques such as thermal analyses and infrared spectroscopy. Stability and solubility tests were also performed on these materials. This work aimed to evaluate the physicochemical properties of these solid forms, which revealed thermal stability improvement. Dissociation of the new phases was observed in water, though. This fact is consistent with the reported observed layered structures and BFDH morphology calculations.

Keywords: diclofenac; nucleobases; mechanochemical synthesis; multicomponent materials; pharmaceutical solid forms

1. Introduction

Diclofenac (DIC), a phenylacetic derivative non-steroidal anti-inflammatory drug (NSAID), is widely used in human and veterinary practice for the treatment of acute and chronic pain as well as in inflammatory and degenerative rheumatic diseases [1,2]. Diclofenac exerts its action through the inhibition of cyclooxygenase-1 (COX-1) and cyclooxygenase-2 (COX-2) enzymes, which inhibit the synthesis of prostaglandins [3]. According to the Biopharmaceutics Classification System (BCS), DIC is a class II drug with low solubility and high permeability [4]. Due to its low solubility (0.9 ± 0.1 µg/mL) [5], achieving its minimum effective concentration requires a higher dosage in the formulation. However, the side effects of DIC have shown dosage-dependency; these include gastrointestinal damage and bleeding, nausea, hepatotoxicity, or renal failure. Moreover, when DIC is administrated orally, its low solubility increases the residence time in the stomach and the contact with the gastric mucosa, increasing the risk of gastric damage [6]. However, poor solubility is a major drawback not only for DIC but also for other Active Pharmaceutical Ingredients (APIs). For that reason, significant efforts have been made by both the industry and academia to develop new methodologies to enhance the physicochemical properties of APIs. Pharmaceutical multicomponent solid forms have gained much interest in the last decade due to their great potential to overcome drug performance limitations [7]. These solid forms are crystalline materials composed of two or more components. At least one must be an API, and the other, called cocrystal former or coformer, must be pharmaceutically acceptable, which means to be recognized as a safe molecule. Both components are in a stoichiometric ratio and interact through non-covalent interactions, mainly hydrogen bonds. These non-covalent interactions guide the organization of the molecules in the crystalline structure and allow the modulation of the physicochemical properties without covalent alterations of the API, whose activity and efficacy remain intact [8]. The literature reports pharmaceutical salts and cocrystals of DIC with amide (isonicotinamide [9]), amine (metformin [10], L-proline [11]), and xanthine (theophylline [12]), as well as pyridine-based coformers [13].

Nucleobases, the main component of nucleic acids, have attracted interest from the crystal engineering point of view because they can establish different hydrogen bond patterns [14]. This ability has been explored previously to form cocrystals and salts through NH···O=C hydrogen bond motifs [15]. Amine-carbonyl synthon has a remarkable key role in the transfer of genetic information and nucleic acid-protein recognition [16]. Moreover, nucleobase-derived drugs exhibit different biological roles, including anti-viral, antibacterial, and antitumoral activities [17,18].

This work reports the synthesis and physicochemical characterization of new multi-component forms with diclofenac and a nucleobase: adenine, cytosine, and isocytosine (Scheme 1). The single crystal structure of all solid forms is thoroughly described, providing valuable insights into the structural differences that drive their physicochemical properties, mainly stability and solubility.

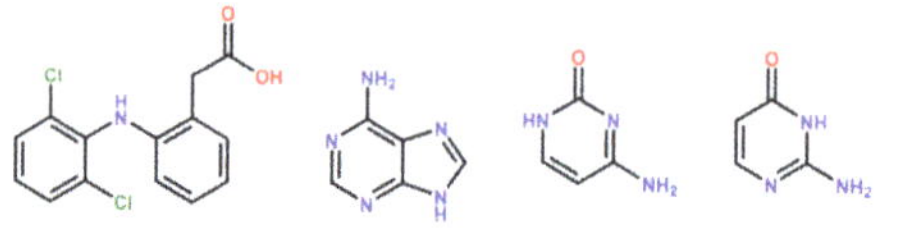

Scheme 1. Chemical formula of diclofenac (DIC), adenine (ADE), cytosine (CYT), and isocytosine (ICT).

2. Materials and Methods

2.1. Materials

Sodium diclofenac (DICNa), adenine, cytosine, and isocytosine were commercially available from Sigma-Aldrich (purity > 98%, Sigma-Aldrich, St. Louis, MO, USA). All solvents were also purchased from Sigma-Aldrich and were used as received.

Synthesis of Diclofenac Acid Form

Diclofenac acid form (DIC) was obtained from hydrolysis of DICNa. For this purpose, 5 mol of DICNa (1.590 g) were dissolved in 30 mL of ultrapure water (Milli-Q, Millipore, Burlington, MA, USA) at 40 °C. HCl 1 M was added dropwise to the solution until no more diclofenac was precipitating. The product was filtrated and washed three times with cold deionized water and let dry at 35 °C for 24 h. Powder X-ray diffraction (PXRD) was used to confirm the purity of DIC.

2.2. Coformer Selection

A search of the Cambridge Structural Database (CSD) [19] was conducted to identify complementary functional groups with the potential for molecular recognition with DIC. A virtual cocrystal screening was performed afterwards using COSMOQuick software [20] (COSMOlogic, Germany, Version 1.4), calculating the excess enthalpy (H_{ex}) of mixing between DIC and selected coformers from an internal library.

2.3. General Procedure for Mechanochemical Synthesis

Mechanochemical experiments were carried out via liquid-assisted grinding (LAG) using a Retsh MM200 ball mill (Retsch, Haan, Germany) operating for 30 min at a 25 Hz frequency using methanol as a solvent.

Synthesis of DIC–ADE: A mixture of DIC (74.1 mg, 0.25 mmol) and ADE (33.80 mg, 0.25 mmol) in a 1:1 stoichiometric ratio was placed in a 10 mL stainless-steel jar along with 100 µL of methanol and two stainless-steel balls of 5 mm diameter.

Synthesis of DIC–CYT: A mixture of DIC (74.1 mg, 0.25 mmol) and CYT (22.20 mg, 0.25 mmol) in a 1:1 stoichiometric ratio was placed in a 10 mL stainless-steel jar along with 100 µL of methanol and two stainless-steel balls of 5 mm diameter.

Synthesis of DIC–ICT: A mixture of DIC (74.1 mg, 0.25 mmol) and ICT (22.20 mg, 0.25 mmol) in a 1:1 stoichiometric ratio was placed in a 10 mL stainless-steel jar along with 100 µL of methanol and two stainless-steel balls of 5 mm diameter.

2.4. Powder X-ray Diffraction (PXRD)

PXRD data were collected using a Bruker D8 Advance Series II Vario diffractometer (Bruker-AXS, Karlsruhe, Germany) equipped with a LYNXEYE detector and Cu-Kα_1 radiation (1.5406 Å). Diffraction patterns were collected over a 2θ range of 5–60° using a continuous step size of 0.02° and a total acquisition time of 30 min.

2.5. Preparation of Single Crystals

Single crystals were grown from saturated solutions (methanol) of the polycrystalline material obtained from LAG synthesis. Suitable single crystals for X-ray diffraction studies were grown by slow solvent evaporation at room temperature for two days.

2.6. Single-Crystal X-ray Diffraction (SCXRD)

Measured crystals were prepared under inert conditions immersed in perfluoropolyether as protecting oil for manipulation. Suitable crystals were mounted on MiTeGen Micromounts™, and these samples were used for data collection. Data for DIC–ADE, DIC–CYT, and DIC–ICT were collected with a Bruker D8 Venture diffractometer with graphite monochromated Cu-Kα radiation (λ = 1.54178 Å). The data were processed with APEX4 suite [21]. The structures were solved by intrinsic phasing using the ShelXT program [22], which revealed the position of all non-hydrogen atoms. These atoms were refined on F^2 by a full-matrix least-squares procedure using an anisotropic displacement parameter [23]. All hydrogen atoms were located in difference Fourier maps and included as fixed contributions riding on attached atoms with isotropic thermal displacement parameters 1.2 or 1.5 times those of the respective atom. The Olex2 software was used as a graphical interface [24]. Intermolecular interactions were calculated using PLATON [25]. Molecular graphics were generated using Mercury [26,27]. The crystallographic data for the reported structures were deposited with the Cambridge Crystallographic Data Center as supplementary publication no. CCDC 2180776-2180778. Additional crystal data are shown in Table 1. Copies of the data can be obtained free of charge at https://www.ccdc.cam.ac.uk/structures/.

Table 1. Crystallographic data and structure refinement of DIC polymorphs and new solid forms.

Compound Name	DIC form I *	DIC form II *	DIC–ADE	DIC–CYT	DIC–ICT
Formula	$C_{14}H_{11}Cl_2NO_2$	$C_{14}H_{11}Cl_2NO_2$	$C_{19}H_{16}Cl_2N_6O_2$	$C_{36}H_{32}Cl_4N_8O_6$	$C_{36}H_{32}Cl_4N_8O_6$
Formula weight	296.14	296.14	431.28	814.49	814.49
Crystal system	Monoclinic	Monoclinic	Triclinic	Orthorhombic	Triclinic
Space group	C2/c	P2$_1$/c	P-1	Pca2$_1$	P1
a/Å	20.226 (4)	8.384 (2)	7.0545 (2)	13.8431 (4)	4.720 (2)
b/Å	6.971 (3)	10.898 (2)	10.3452 (4)	8.4502 (4)	9.701 (3)
c/Å	20.061 (4)	14.822 (5)	14.3310 (5)	32.0448 (11)	20.189 (7)
α/°	90	90	97.913 (2)	90	84.328 (16)
β/°	109.64 (2)	92.76 (2)	104.237 (2)	90	88.058 (16)
γ/°	90	90	100.934 (2)	90	85.963 (16)
V/Å^3	2664 (1)	1352.7 (6)	976.51 (6)	3748.5 (2)	917.4 (6)
Z	8	4	2	4	1
Dc/g cm^{-3}	1.477	1.454	1.467	1.443	1.474
F(000)	1216	608	444	1680	420

Table 1. Cont.

Reflections collected	4383	4079	12246	29559	6125
Unique reflections	2589	3940	3398	6581	6125
Data/restraints/parameters	2582/36/217	3937/36/216	3398/0/263	6581/1/487	6125/3/488
Goodness-of-fit (on F^2)	1.057	1.005	1.066	1.016	1.059
$R1$ [$I > 2\sigma(I)$]	0.0374	0.0397	0.0526	0.0506	0.0683
$wR2$ [$I > 2\sigma(I)$]	0.0992	0.0859	0.1531	0.1302	0.1800
Absolute structure parameter	-	-	-	0.067 (15)	0.01 (2)
CCDC	128772	128771	2180776	2180777	2180778

* Reported in [28].

2.7. Thermal Analysis

Simultaneous thermogravimetric analysis (TGA) and differential scanning calorimetry (DSC) measurements were performed using a Mettler Toledo TGA/DSC1 thermal analyzer (Mettler Toledo, Columbus, OH, USA). Samples (3–5 mg) were placed into sealed aluminium pans and heated in a stream of nitrogen (100 mL min^{-1}) from 25 to 400 °C at a heating rate of 10 °C min^{-1}.

2.8. Infrared Spectroscopy

Fourier-transform infrared (FT–IR) spectroscopic measurements were performed on a Bruker Tensor 27 FT–IR instrument (Bruker Corporation, Billerica, MA, USA) equipped with a single-reflection diamond crystal platinum ATR unit and OPUS data collection program. The scanning range was from 4000 to 400 cm^{-1} with a resolution of 4 cm^{-1}.

2.9. Stability Test

Stability in aqueous solution was evaluated through slurry experiments. Excess of powder samples of each phase was added to 1 mL of water and stirred for 24 h in sealed vials. The solids were collected, filtered, and dried for further analysis by PXRD.

Stability at accelerated ageing conditions was also studied: 200 mg of solid was placed in watch glasses and left at 40 °C in 75% relative humidity using a Memmert HPP110 climate chamber (Memmert, Schwabach, Germany). The samples were subjected to the above-accelerated stability conditions for two months. PXRD was used to monitor the stability of the solid forms.

2.10. Solubility Test

Solubility studies were performed using the Crystal16 equipment (Technobis Crystallization Systems, Alkmaar, The Netherlands) in water. The equipment comprised four individually controlled reactors, each with a working volume of 1 mL, allowing the measurement of cloud and clear points based on the turbidity of 16 aliquots of 1 mL of solution in parallel and automatically. Each solution was heated at 0.3 °C/min from 20 to 90 °C with a magnetic stirring rate of 700 rpm, held at this temperature for 10 min and then cooled to 20 °C at 0.3 °C/min. The dissolution temperature for each compound was measured using different amounts of solid, and the solubility data of the pure components were fitted to a quadratic equation [29] using the CrystalClear software (Technobis Crystallization Systems, Alkmaar, The Netherlands).

3. Results and Discussion

3.1. Coformer Selection

Before the experimental trials were conducted, a virtual cocrystal screening was performed to improve the success ratio. A survey on the Cambridge Structural Database (CSD version 5.43, update from June 2022) based on DIC resulted in 70 hits. After excluding the three reported polymorphs [28,30,31] and metal complexes [32–41], the dataset contained 28 hits corresponding to multicomponent forms (salts, cocrystals, hydrates, and solvates). Only in one hit, the dimer DIC–DIC, observed in the monoclinic DIC polymorphs [28,30], was maintained; meanwhile, the remaining hits exhibited common structural features for DIC salts, COO$^-$···amine and COO$^-$··· ammonium synthons, or cocrystals: COOH···N(pyridine) and COOH···N(imidazole) synthons. According to the above-mentioned, our main prerequisite for the coformer selection was having the above-referred N-groups and being a safe molecule. From our library of coformers, two groups of molecules fulfil these criteria: amino acids and nucleobases. COSMOQuick software was used to validate our selection. Table 2 shows calculations from a list of candidates to form multicomponent crystals with DIC. The list includes our reported coformers and other coformer molecule involved in the formation of cocrystals/salts reported in the survey. Compounds with negative H$_{ex}$ values show an increased probability of forming cocrystals.

Table 2. Ranking of potential DIC coformers used in this work, based on COSMOQuick calculations.

Coformer	H$_{ex}$ (kcal/mol)	
Glycine	−5.070	
Proline	−4.743	Ref. [11]
Alanine	−3.949	
Glutamic Acid	−3.699	
Aspartic Acid	−3.285	
Cytosine	**−3.177**	This work
Adenine	**−2.393**	This work
Cysteine	−2.015	
Thymine	−1.498	
Phenylglycine	−1.09	
Isocytosine	**−1.075**	This work

3.2. Mechanochemical Synthesis

Liquid-Assisted Grinding is a versatile and efficient methodology widely used to obtain pharmaceutical multicomponent solid forms [42]. A screening through LAG was conducted with DIC and all coformers listed in Table 2. Unfortunately, despite the promising results obtained by the COSMOQuick analysis, only those LAG synthesis with the coformers ADE, CYT, and ICT were successful and achieved new phases whilst the other coformers yielded only physical mixtures of the two components. The product of these reactions was characterized by PXRD and compared with the X-ray powder pattern of the parent components (Table S1, Figure S1). Only experiments using ADE, CYT, and ICT as coformers provided a new PXRD pattern and were used for subsequent characterization (Figure 1). After the screening procedure, the work was focused on the search for fine-tuning conditions to obtain multicomponent materials of DIC with ADE, CYT, and ICT. Neat grinding experiments resulted in physical mixtures of the components (Figure S1). Neat grinding approach only led to physical mixtures (Figure S2). It is reported that this synthesis technique sometimes yields products with low crystallinity, partial reactions, or not even a reaction at all [43]. However, it is well known that adding small amounts of liquids accelerate the reaction, which essentially drove us to the idea of using LAG synthesis instead. LAG experiments were then performed using methanol and different stoichiometries (1:1, 1:2, and 2:1) (Figure S3). A new common pattern was observed in the three stoichiometries. However, in the 1:2 and 2:1 ratios, there were also peaks corresponding to the coformer and DIC, respectively. Only the 1:1 ratio provided unique different PXRD patterns. Comparing these patterns with those simulated from the crystal structures confirmed the monophasic nature of the bulk solids (Figure S4).

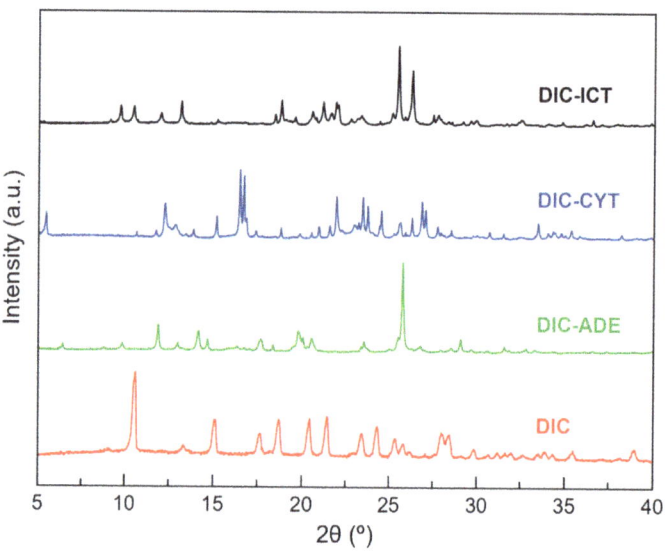

Figure 1. PXRD patterns of DIC, DIC–ADE, DIC–CYT, and DIC–ICT obtained by Liquid-Assisted Grinding (LAG) with methanol in a 1:1 ratio.

3.3. Structural Analysis of Multicomponent Forms

DIC–ADE cocrystal crystallized in the triclinic P-1 space group. The asymmetric unit is composed of DIC and ADE in a 1:1 stoichiometric ratio, where ADE adopts its most stable and expected 9H tautomeric form. DIC exhibits its aromatic ring twisted out (dihedral angle: 71.92°) and is stabilized by an intramolecular N–H(amine)···O(carbonyl) hydrogen bond (Figure 2a). Centrosymmetric H-bonded adenine dimers (N29-H29···N23#2 2.00 Å, 166.6°; #2: -x + 3,-y + 2,-z) aggregate through the Hoogsteen edge (N26-H26B···N27#1: 2.10 Å, 162.7°; #1:-x + 2,-y + 1,-z), creating infinite zig-zag chains. DIC molecules connect to the chain structure by H-bonding interactions through the Watson–Crick edge (O2-H2···N21: 1.84 Å, 169°; N26-H26A···O1: 2.10 Å, 169°), resulting in an infinite tape structure (Figure 2b). C–H···F hydrogen bonds reinforce this structure and also connect adjacent tapes. Finally, the 3D structure is accomplished by piling these tapes through C=O···π and C-H···π interactions among DIC and aromatic rings from the adenine (Figure 2c).

DIC–CYT crystallized as a molecular salt in the orthorhombic Pca2$_1$ spacegroup. The asymmetric unit consisted of two symmetry-independent molecules of diclofenac anion and two symmetry-independent molecules of cytosinium cation in a 1:1 stoichiometric ratio (Figure 3a). Cytosinium over hemicytosinium duplex formation was observed in agreement with the cutoff pK$_a$ value for acids reported by Sun et al. [44] (pK$_a$ value for DIC: 4.15). The analysis of the C–O bond distances of the carboxylate group of DIC supports the salt formation [45]. In the DIC–CYT system, C–O distances were indicative of a deprotonated acid, as expected for a salt with ΔD_{C-O} values of 0.001 Å and 0.002 Å for both DIC anions, respectively, according to the ΔD_{C-O} values observed in salts (typically less than 0.03 Å). As in DIC–ADE, the two aromatic rings of diclofenac are bent out of plane, with dihedral angles of 81.78 and 84.06°. In the crystal, DIC$^-$ and CYT–H$^+$ form an alternating layered structure where cytosinium molecules are associated through single-point N–H···O bonds, graph set $C_1^1(6)$, generating CYT–H$^+$···CYT–H$^+$ chains running along the a-axis [Figure 3b]. DIC$^-$ layers are reinforced by C–H···F hydrogen bonds. The two-point 2-amino-pyridinium–carboxylate synthon (N4A–H4AA···O1A, 1.85 Å, 178°, N2A–H2A$^+$···O2A, 1.92 Å, 176.3°, and N4B–H4BA···O1B, 1.84 Å, 178.4°, N2B–H2B$^+$···O2B, 1.92 Å, 176.5°) associates the DIC$^-$ and CYT–H$^+$ layers, generating the supramolecular 3D structure.

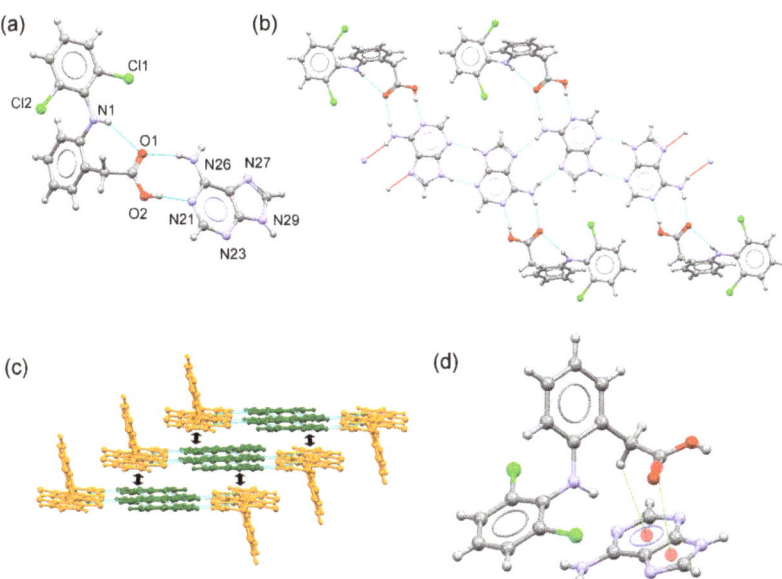

Figure 2. (**a**) Asymmetric unit of the DIC–ADE cocrystal. (**b**) Fragment of the tape structure generated by H-bonding interactions. (**c**) Detailed view of the crystal packing of the DIC–ADE cocrystal. Orange: DIC molecules, green: ADE molecules. (**d**) C=O···π and C-H···π interactions in the DIC–ADE cocrystal.

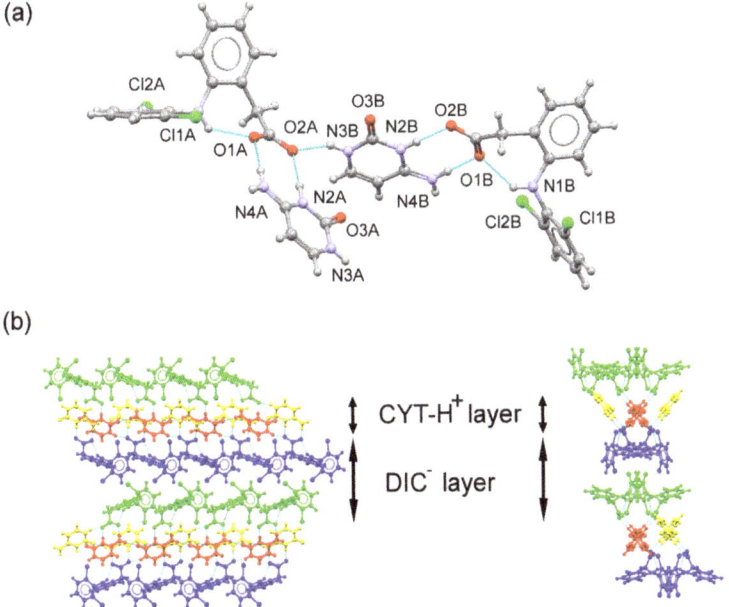

Figure 3. (**a**) Asymmetric unit of the DIC–CYT molecular salt. (**b**) Detailed view of the packing arrangement of DIC$^-$ anions (blue and green) and CYT–H$^+$ cations (red and yellow) in the DIC–CYT crystal structure (viewed along the b and a axes), showing an alternating layered structure.

DIC–ICT crystallized in the triclinic P1 spacegroup. Both diclofenac and isocytosine components are present in their neutral and ionic forms, resulting in a hybrid solid with a cocrystal and a salt in the asymmetric unit (Figure 4a). In the DIC–ICT system, C–O distances confirmed the presence of carboxylate and carboxylic groups, as expected for ΔD_{C-O} values observed for the two symmetry-independent diclofenac molecules (0.080 Å and 0.006 Å for neutral diclofenac and diclofenac anion, respectively). Isocytosine and isocytosinium molecules formed a dimeric structure through H-bonds involving the 2-amino-pyridinium–carbonyl synthon (graph set motif $D_1^1(2)$). These dimers connect with adjacent dimers using the amine-carbonyl synthon (graph set motif $D_1^1(2)$), generating a chain structure. The two-point 2-amino-pyridine–carboxylic (N4B–H4BA···O1B, 2.02 Å, 166.5°, and O2B–H2B···N2B#1, 1.80 Å, 171.7°; #1: x + 1, y, z) and 2-amino-pyridinium–carboxylate (N4A–H4AB···O1A, 1.82 Å, 158.1°, and N2A–H2A$^+$···O2A, 1.99 Å, 174.0°) synthons connect components to build up a ribbon structure (Figure 4b). C–H···π interactions (C13B-H13D···Cg; H···Cg distance: 2.97 Å; C-H···Cg angle: 118°; Cg = C7B-C8B-C9B-C10B-C11B-C12B) associate these ribbons to form a layered structure. Finally, weak C–H···F hydrogen bonds connect these layers to create the 3D structure.

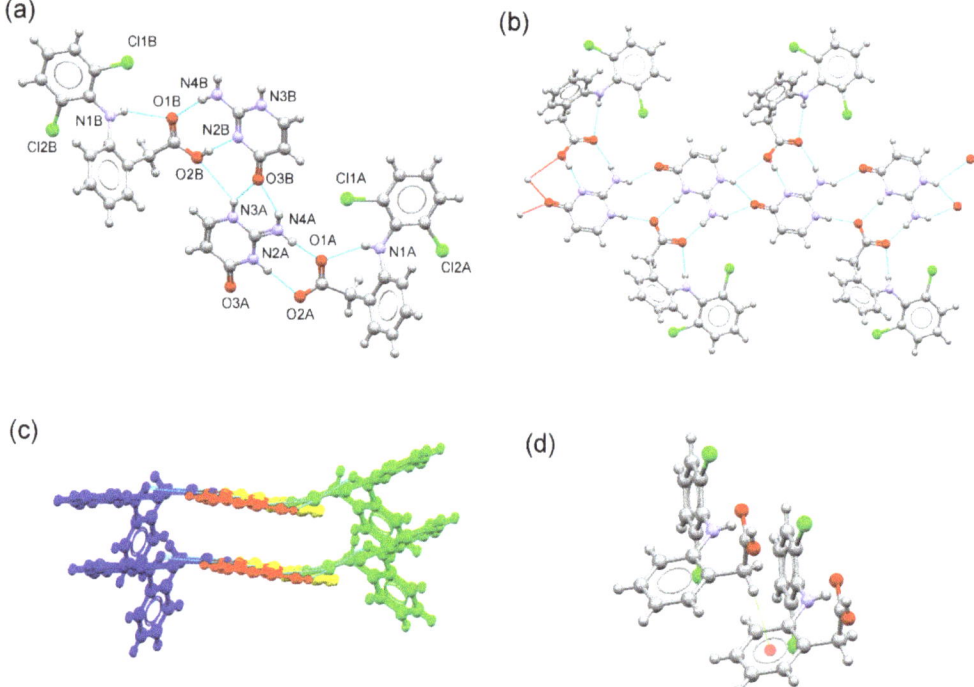

Figure 4. (**a**) Asymmetric unit of the DIC–ICT multicomponent form. (**b**) Fragment of the ribbon structure generated by H-bonding interactions between DIC components and a chain of –ICT–H$^+$···ICT– dimmers. (**c**) Detailed view of the packing arrangement of ribbons structures containing DIC (blue), DIC$^-$ (green), ICT (yellow), and ICT–H$^+$ (red), building up a layered structure. (**d**) Detailed view of the C–H···π interaction between DIC molecules.

3.4. Thermal Analysis

DSC was used to evaluate the thermal behaviour and to determine the melting point of the new DIC phases. Figure 5 shows the melting point of DIC, as well as the DSC traces of DIC–ADE, DIC–CYT, and DIC–ICT. Each plot shows a well-defined endothermic event that corresponds with the melting point of the material. A single endothermic transition indicates the absence of solvation or hydration phenomena and also demonstrates the stability of the phase until the melting point. Above the melting point, some endothermic events are also observed, corresponding to the degradation of the samples. The multicomponent materials display a melting point that falls in a region between the melting point of DIC (179 °C) and the coformer (ADE: 360 °C; CYT: 320–325 °C; ICT:248–254 °C). This feature has already been described by other researchers [46]. A higher melting point was obtained through multicomponent crystallization, resulting in better thermal stability, probably due to stronger intermolecular interactions between DIC and nucleobases. TGA showed no weight loss until melting, suggesting that the new DIC phases were not hydrated or solvated. The occurrence of mass loss was observed after melting points, which was attributed to the degradation of cocrystals (Figure S5).

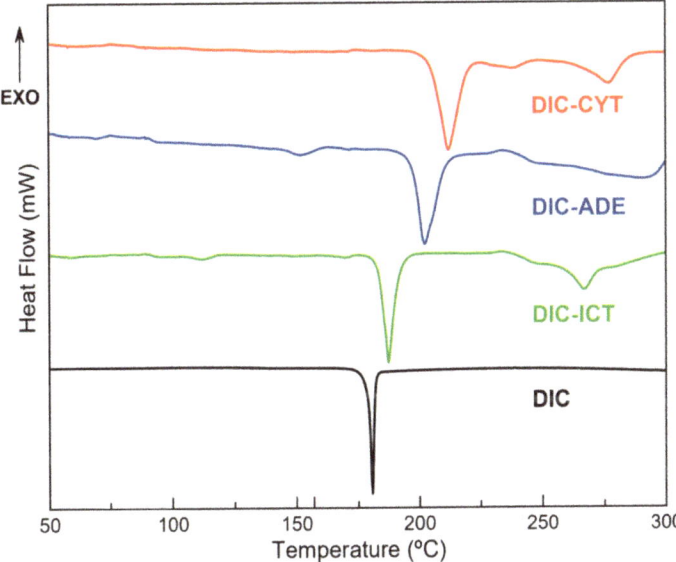

Figure 5. Differential scanning calorimetry (DSC) traces of DIC and multicomponent compounds DIC–ADE, DIC–CYT, and DIC–ICT.

3.5. Fourier Transform Infrared (FT-IR) Spectroscopy

Due to its simplicity and reduced consumption of time and samples, FT–IR spectroscopy is a widely used technique for detecting new multicomponent materials [47]. Functional groups exhibit defined bands in the IR spectrum, and intermolecular interactions, such as hydrogen bonds, induce changes in the position of these bands. Hence, the study of the shifts can detect the formation of a cocrystal or a salt and gives information about the groups involved in the interaction [48].

Figure 6 shows the FT–IR spectra of DIC and DIC multicomponent materials. The DIC spectrum has a characteristic band at 3322 cm^{-1}, ascribed to the stretching mode of –NH. In DIC–ADE, DIC–CYT, and DIC–ICT, this band is shifted to 3326, 3300, and 3298 cm^{-1}, respectively. Another characteristic band of DIC is the C=O stretching vibration that appears at 1961 cm^{-1}. This band is shifted to 1671 (DIC–ADE), 1693 (DIC–CYT), and 1678 cm^{-1} (DIC–ICT). FT–IR data support the information observed in SCXRD, where –COOH and –NH groups from DIC and coformers, respectively, drive the formation of the crystalline structures.

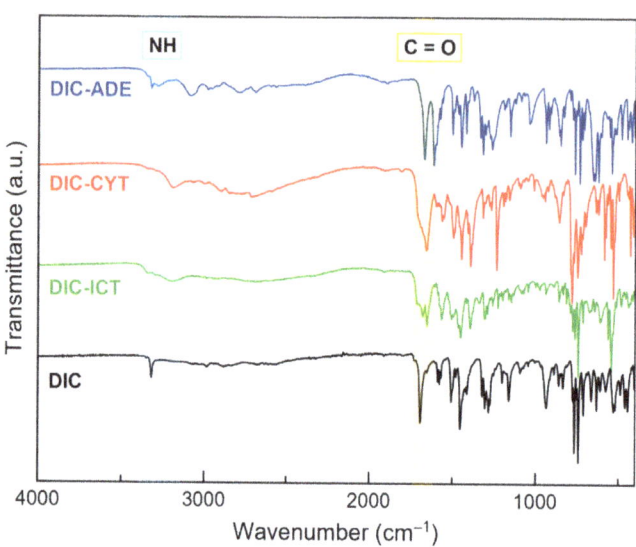

Figure 6. Fourier transform infrared (FT–IR) spectra of DIC, DIC–ADE, DIC–CYT, and DIC–ICT.

3.6. Stability Studies

Thermodynamic stability in an aqueous solution was evaluated by placing an excess of the sample in a vial and stirring in deionized water at 25 °C. After 24 h, the product was filtered, dried at room temperature, and characterized by PXRD. Powder patterns showed high stability for the pure DIC and DIC–ADE cocrystal. However, DIC–CYT and DIC–ICT phases could not remain stable for more than 3 h and 30 min, respectively (Figures 7 and S7). Multicomponent DIC phases were also stored at accelerating ageing conditions (40 °C, 75% RH). At the determined time, samples were characterized by PXRD to assess their stability. DIC was included in the experiment for better comparison with the new materials. Under these conditions, it was observed that all samples remained stable for two months (Figure 8).

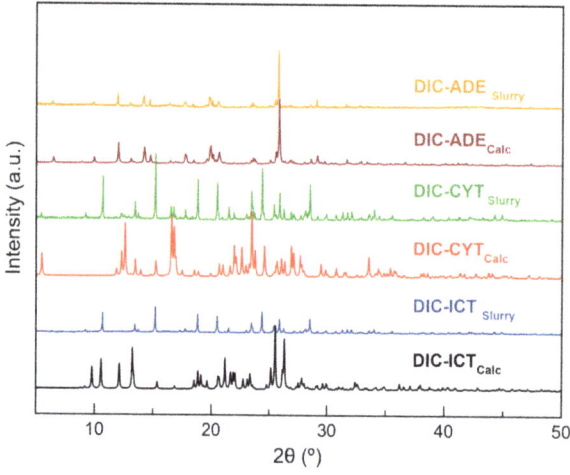

Figure 7. PXRD diagrams corresponding to the stability of DIC–ADE, DIC–CYT, and DIC–ICT in aqueous slurry experiments at 24 h.

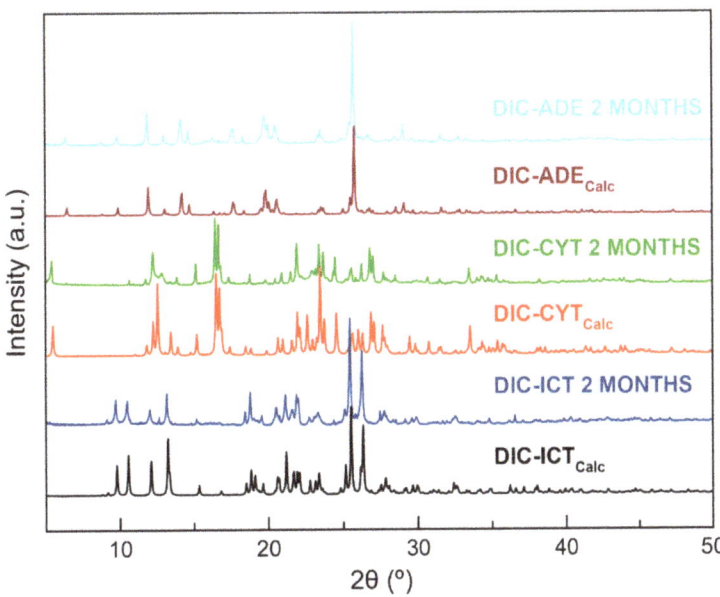

Figure 8. PXRD diagrams corresponding to the stability of DIC–ADE, DIC–CYT, and DIC–ICT in accelerated ageing conditions (40 °C, 75% RH) at two months.

3.7. Solubility Studies

As observed in the previous section, only the DIC–ADE cocrystal was thermodynamically stable in a water solution at room temperature. Initial attempts to determine solubility by the shake-flask method [49] were not possible due to the overlap of UV absorption maxima of both the API and coformer (Figure S8). Evaluation of solubility was then performed by a polythermal method using the Crystal16 equipment. Results showed an improvement in the solubility of DIC–ADE (0.993 mg/mL, Figure 9) compared with the reported solubility of DIC (0.9 μg/mL) [5]. Although the difference in solubility between DIC and the DIC–ADE cocrystal was significant, the amount of solubility improvement is not significant compared with the solubility of the sodium salt (16.18 mg/mL) [50]. The layered structure observed in the DIC–ADE cocrystal directly impacts the solubility improvement of DIC. Different studies have reported the effect of a layer structure consisting of high-solubility molecules on the solubility of multicomponent solid forms [51–53]. Although the dimeric DIC structure is disrupted and a better solubility is obtained in DIC–ADE, the intercalated layers composed of low water-soluble ADE molecules do not confer enough solubility improvement themselves in comparison with other multicomponent DIC solids.

A potential risk observed in the use of multicomponent systems is their tendency to experience unexpected dissociation in contact with water or with high relative humidity (RH), which leads to a return to the respective free API and coformer [54,55] and denies the solubility advantage achieved by multicomponent solid formation. To rationalize the dissociation observed for DIC–CYT and DIC–ICT solids, crystal morphologies of the three reported multicomponent DIC forms were computed using the Bravais–Friedel–Donnay–Harker (BFDH) method included in the visualization software package Mercury [27]. As described previously, all the crystal structures consist of alternate layers of type DIC···coformer···DIC··· and these supramolecular arrangements seem responsible for the enhanced properties. Figure 10 shows the predicted morphologies for the reported multicomponent solids. Notably, the facets with the largest surface, following the order: DIC–ICT (57.2%) > DIC–CYT (50%) > DIC–ADE (24.2%), contain hydrogen bond donor and acceptor groups that potentially could interact with water during the dissolution process. Water solubility depends not only on the groups exposed on the surface of a crystal but on other different factors, including density, coformer solubility, or lattice energy [56]. We cannot argue that the high polarity of the crystal surfaces could impact the solubility performance by itself, but it does affect the dissolution of the reported solids as evidenced by the rapid dissociation observed for DIC–CYT and DIC–ICT during the slurry experiments in water. Dissolution is carried out at higher rate in the DIC–CYT and DIC–ICT species. As expected, dissolution in these species is favored by extensive surface exposure. On the other hand, dissolution of the DIC–ADE phase occurs at a slower rate, evidenced by the apparent stability at 24 h.

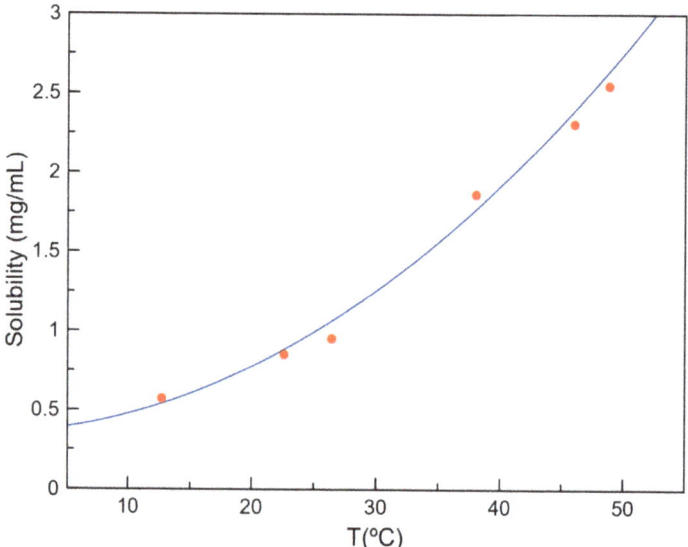

Figure 9. Solubility curve for DIC–ADE in water as a function of concentration and temperature.

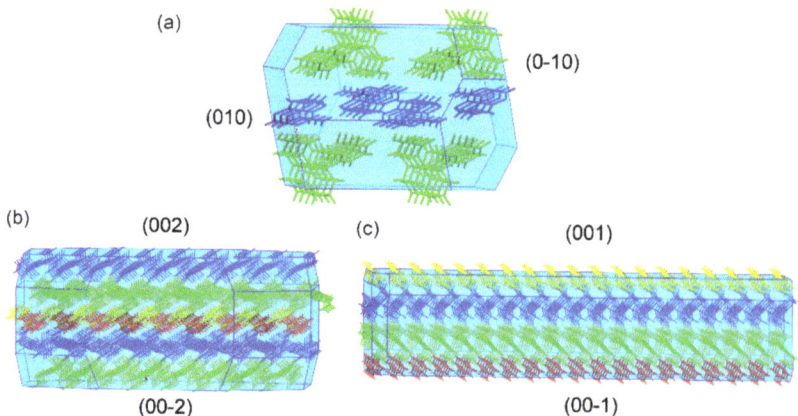

Figure 10. BFDH-predicted morphologies of (**a**) DIC–ADE I (green: DIC, blue: ADE); (**b**) DIC–CYT (blue and green: DIC, yellow and red: CYT), and (**c**) DIC–ICT (DIC: blue and green: ICT: red and yellow), showing the largest faces.

4. Conclusions

In conclusion, we have described three new pharmaceutical multicomponent crystals containing DIC and nucleobases (ADE, CYT, and ICT) as coformers. Expected heterosynthons assist formation of the new solid forms, disrupting the robust acid:acid dimmer synthon observed in reported DIC polymorphs. All solids consist of alternated layered structures connected by hydrogen bonds. This supramolecular organization confers good thermal stability, and good stability under accelerated ageing conditions and seems to have an important role in the dissolution properties of the solids. Relevant insights are inferred from the BFDH calculations where CYT and ICT, containing solids, possess a large crystal surface that expose hydrogen donor and acceptor groups, which interact with the water molecules of the bulk solvent.

Supplementary Materials: The following supporting information can be downloaded at: https://www.mdpi.com/article/10.3390/cryst12081038/s1, Figure S1. PXRD patterns of the LAG screening experiments with the coformers of Table 2; Figure S2. PXRD patterns of DIC—ADE, DIC—CYT, and DIC—ICT after neat grinding; Figure S3. PXRD patterns of DIC—ADE, DIC—CYT, and DIC—ICT after LAG in methanol using different stoichiometries; Figure S4. Experimental PXRD pattern of DIC—ADE, DIC—CYT, and DIC—ICT, compared with DIC, coformers, and the corresponding calculated powder patterns; Figure S5. TGA traces of DIC–ADE (top), DIC—CYT (middle), and DIC–ICT (bottom); Figure S6. PXRD patterns of DIC–ADE (top), DIC—CYT (middle), and DIC–ICT (bottom) with respect to the stability under accelerated ageing conditions (40 °C, 75% RH) at different time intervals; Figure S7. PXRD patterns of DIC–ADE (top), DIC—CYT (middle), and DIC–ICT (bottom) after the stability slurry assay (at 25 °C, during 24 h, in water); Figure S8. Overlapping UV spectra of diclofenac (DIC) and nucleobase coformers (ADE: adenine, CYT: cytosine, ICT: isocytosine); Figure S9. ORTEP representation showing the asymmetric unit of DIC—ADE with an atom numbering scheme (thermal ellipsoids are plotted with the 50% probability level); Figure S10. ORTEP representation showing the asymmetric unit of DIC—CYT with atom numbering scheme (thermal ellipsoids are plotted with the 50% probability level); Figure S11. ORTEP representation showing the asymmetric unit of DIC—ICT with atom numbering scheme (thermal ellipsoids are plotted with the 50% probability level); Table S1. Results of the LAG experiments between DIC and selected coformers; Table S2. Hydrogen bonds for DIC—ADE [Å and deg.]; Table S3. Hydrogen bonds for DIC—CYT [Å and deg.]; Table S4. Hydrogen bonds for DIC—ICT [Å and deg.].

Author Contributions: Conceptualization and methodology, D.C.-L.; formal analysis and investigation, C.A.-P., F.J.A.-M., H.M.B.-R., J.N.-G., A.D.-M.; writing—original draft preparation, D.C.-L., F.J.A.-M.; writing—review and editing, D.C.-L.; funding acquisition, A.D.-M. and D.C.-L.; supervision, D.C.-L. All authors have read and agreed to the published version of the manuscript.

Funding: This research was funded by Spanish Agencia Estatal de Investigación of the Ministerio de Ciencia, Innovación y Universidades (MICIU) and co-funded with FEDER, UE, project no. PGC2018-102047-B-I00 (MCIU/AEI/FEDER, UE) and project B-FQM-478-UGR20 (FEDER-Universidad de Granada, Spain).

Institutional Review Board Statement: Not applicable.

Informed Consent Statement: Not applicable.

Data Availability Statement: Not applicable.

Acknowledgments: F.J.A.-M. acknowledges an FPI grant (ref. PRE2019-088832).

Conflicts of Interest: The authors declare no conflict of interest.

References

1. Davies, N.M.; Andersen, K.E. Clinical Pharmacokinetics of Diclofenac. *Clin. Pharmacokinet.* **2012**, *33*, 184–213. [CrossRef] [PubMed]
2. Brogden, R.N.; Heel, R.C.; Pakes, G.E.; Speight, T.M.; Avery, G.S. Diclofenac Sodium: A Review of Its Pharmacological Properties and Therapeutic Use in Rheumatic Diseases and Pain of Varying Origin. *Drugs* **2012**, *20*, 24–48. [CrossRef] [PubMed]
3. Gan, T.J. Diclofenac: An Update on Its Mechanism of Action and Safety Profile. *Curr. Med. Res. Opin.* **2010**, *26*, 1715–1731. [CrossRef] [PubMed]

4. Takagi, T.; Ramachandran, C.; Bermejo, M.; Yamashita, S.; Yu, L.X.; Amidon, G.L. A Provisional Biopharmaceutical Classification of the Top 200 Oral Drug Products in the United States, Great Britain, Spain, and Japan. *Mol. Pharm.* **2006**, *3*, 631–643. [CrossRef]
5. Stuart, M.; Box, K. Chasing Equilibrium: Measuring the Intrinsic Solubility of Weak Acids and Bases. *Anal. Chem.* **2005**, *77*, 983–990. [CrossRef]
6. Gomaa, S. Adverse Effects Induced by Diclofenac, Ibuprofen, and Paracetamol Toxicity on Immunological and Biochemical Parameters in Swiss Albino Mice. *J. Basic Appl. Zool.* **2018**, *79*, 5. [CrossRef]
7. Berry, D.J.; Steed, J.W. Pharmaceutical Cocrystals, Salts and Multicomponent Systems; Intermolecular Interactions and Property Based Design. *Adv. Drug Deliv. Rev.* **2017**, *117*, 3–24. [CrossRef]
8. Bolla, G.; Nangia, A. Pharmaceutical Cocrystals: Walking the Talk. *Chem. Commun.* **2016**, *52*, 8342–8360. [CrossRef]
9. Báthori, N.B.; Lemmerer, A.; Venter, G.A.; Bourne, S.A.; Caira, M.R. Pharmaceutical Co-Crystals with Isonicotinamide-Vitamin B3, Clofibric Acid, and Diclofenac-and Two Isonicotinamide Hydrates. *Cryst. Growth Des.* **2011**, *11*, 75–87. [CrossRef]
10. Feng, W.Q.; Wang, L.Y.; Gao, J.; Zhao, M.Y.; Li, Y.T.; Wu, Z.Y.; Yan, C.W. Solid State and Solubility Study of a Potential Anticancer Drug-Drug Molecular Salt of Diclofenac and Metformin. *J. Mol. Struct.* **2021**, *1234*, 130166. [CrossRef]
11. Nugrahani, I.; Utami, D.; Ibrahim, S.; Nugraha, Y.P.; Uekusa, H. Zwitterionic Cocrystal of Diclofenac and L-Proline: Structure Determination, Solubility, Kinetics of Cocrystallization, and Stability Study. *Eur. J. Pharm. Sci.* **2018**, *117*, 168–176. [CrossRef]
12. Surov, A.O.; Voronin, A.P.; Manin, A.N.; Manin, N.G.; Kuzmina, L.G.; Churakov, A.V.; Perlovich, G.L. Pharmaceutical Cocrystals of Diflunisal and Diclofenac with Theophylline. *Mol. Pharm.* **2014**, *11*, 3707–3715. [CrossRef]
13. Goswami, P.K.; Kumar, V.; Ramanan, A. Multicomponent Solids of Diclofenac with Pyridine Based Coformers. *J. Mol. Struct.* **2020**, *1210*, 128066. [CrossRef]
14. Sivakova, S.; Rowan, S.J. Nucleobases as Supramolecular Motifs. *Chem. Soc. Rev.* **2005**, *34*, 9–21. [CrossRef]
15. Koch, E.S.; McKenna, K.A.; Kim, H.J.; Young, V.G.; Swift, J.A. Thymine Cocrystals Based on DNA-Inspired Binding Motifs. *CrystEngComm* **2017**, *19*, 5679–5685. [CrossRef]
16. Sarai, A.; Kono, H. Protein-DNA Recognition Patterns and Predictions. *Annu. Rev. Biophys. Biomol. Struct.* **2005**, *34*, 379–398. [CrossRef]
17. Xia, Y.; Qu, F.; Peng, L. Triazole Nucleoside Derivatives Bearing Aryl Functionalities on the Nucleobases Show Antiviral and Anticancer Activity. *Mini-Rev. Med. Chem.* **2010**, *10*, 806–821. [CrossRef]
18. Sun, R.; Wang, L. Inhibition of Mycoplasma Pneumoniae Growth by FDA-Approved Anticancer and Antiviral Nucleoside and Nucleobase Analogs. *BMC Microbiol.* **2013**, *13*, 184. [CrossRef]
19. Allen, F.H. The Cambridge Structural Database: A Quarter of a Million Crystal Structures and Rising. *Acta Crystallogr. Sect. B Struct. Sci.* **2002**, *58*, 380–388. [CrossRef]
20. Loschen, C.; Klamt, A. Solubility Prediction, Solvate and Cocrystal Screening as Tools for Rational Crystal Engineering. *J. Pharm. Pharmacol.* **2015**, *67*, 803–811. [CrossRef]
21. Bruker APEX4, APEX4 V2021.1 2021; Bruker-AXS: Madison, WI, USA, 2021.
22. Sheldrick, G.M. SHELXT—Integrated Space-Group and Crystal-Structure Determination. *Acta Crystallogr. Sect. A Found. Crystallogr.* **2015**, *71*, 3–8. [CrossRef] [PubMed]
23. Sheldrick, G.M. Crystal Structure Refinement with SHELXL. *Acta Crystallogr. Sect. C Struct. Chem.* **2015**, *71*, 3–8. [CrossRef] [PubMed]
24. Dolomanov, O.V.; Bourhis, L.J.; Gildea, R.J.; Howard, J.A.K.; Puschmann, H. OLEX2: A Complete Structure Solution, Refinement and Analysis Program. *J. Appl. Crystallogr.* **2009**, *42*, 339–341. [CrossRef]
25. Spek, A.L. Structure Validation in Chemical Crystallography. *Acta Crystallogr. Sect. D Biol. Crystallogr.* **2009**, *65*, 148–155. [CrossRef]
26. Macrae, C.F.; Bruno, I.J.; Chisholm, J.A.; Edgington, P.R.; McCabe, P.; Pidcock, E.; Rodriguez-Monge, L.; Taylor, R.; Van De Streek, J.; Wood, P.A. Mercury CSD 2.0—New Features for the Visualization and Investigation of Crystal Structures. *J. Appl. Crystallogr.* **2008**, *41*, 466–470. [CrossRef]
27. MacRae, C.F.; Sovago, I.; Cottrell, S.J.; Galek, P.T.A.; McCabe, P.; Pidcock, E.; Platings, M.; Shields, G.P.; Stevens, J.S.; Towler, M.; et al. Mercury 4.0: From Visualization to Analysis, Design and Prediction. *J. Appl. Crystallogr.* **2020**, *53*, 226–235. [CrossRef]
28. Castellari, C.; Ottani, S. Two Monoclinic Forms of Diclofenac Acid. *Acta Crystallogr. Sect. C Cryst. Struct. Commun.* **1997**, *53*, 794–797. [CrossRef]
29. Horst, J.H.T.; Deij, M.A.; Cains, P.W. Discovering New Co-Crystals. *Cryst. Growth Des.* **2009**, *9*, 1531–1537. [CrossRef]
30. Moser, P.; Sallmann, A.; Wiesenberg, I. Synthesis and Quantitative Structure-Activity Relationships of Diclofenac Analogues. *J. Med. Chem.* **1990**, *33*, 2358–2368. [CrossRef]
31. Jaiboon, N.; Yos-In, K.; Ruangchaithaweesuk, S.; Chaichit, N.; Thutivoranath, R.; Siritaedmukul, K.; Hannongbua, S. New Orthorhombic Form of 2-[(2,6-Dichlorophenyl)Amino]Benzeneacetic Acid (Diclofenac Acid). *Anal. Sci.* **2001**, *17*, 1465–1466. [CrossRef]
32. Hamamci Alisir, S.; Dege, N. Crystal Structure of a Mixed-Ligand Silver(I) Complex of the Non-Steroidal Anti-Inflammatory Drug Diclofenac and Pyrimidine. *Acta Crystallogr. Sect. E Crystallogr. Commun.* **2016**, *72*, 1475–1479. [CrossRef] [PubMed]
33. Caglar, S.; Dilek, E.; Caglar, B.; Adiguzel, E.; Temel, E.; Buyukgungor, O.; Tabak, A. New Metal Complexes with Diclofenac Containing 2-Pyridineethanol or 2-Pyridinepropanol: Synthesis, Structural, Spectroscopic, Thermal Properties, Catechol Oxidase and Carbonic Anhydrase Activities. *J. Coord. Chem.* **2016**, *69*, 3321–3335. [CrossRef]

34. García-García, A.; Méndez-Arriaga, J.M.; Martín-Escolano, R.; Cepeda, J.; Gómez-Ruiz, S.; Salinas-Castillo, A.; Seco, J.M.; Sánchez-Moreno, M.; Choquesillo-Lazarte, D.; Ruiz-Muelle, A.B.; et al. In Vitro Evaluation of Leishmanicidal Properties of a New Family of Monodimensional Coordination Polymers Based on Diclofenac Ligand. *Polyhedron* **2020**, *184*, 114570. [CrossRef]
35. Sharma, R.; Sharma, R.P.; Bala, R.; Kariuki, B.M. Second Sphere Coordination in Oxoanion Binding: Synthesis, Spectroscopic Characterisation and Crystal Structures of Trans-[Bis(Ethylenediamine)Dinitrocobalt(III)] Diclofenac and Chlorate. *J. Mol. Struct.* **2007**, *826*, 177–184. [CrossRef]
36. Biswas, P.; Dastidar, P. Anchoring Drugs to a Zinc(II) Coordination Polymer Network: Exploiting Structural Rationale toward the Design of Metallogels for Drug-Delivery Applications. *Inorg. Chem.* **2021**, *60*, 3218–3231. [CrossRef]
37. Lu, C.; Laws, K.; Eskandari, A.; Suntharalingam, K. A Reactive Oxygen Species-Generating, Cyclooxygenase-2 Inhibiting, Cancer Stem Cell-Potent Tetranuclear Copper(II) Cluster. *Dalton Trans.* **2017**, *46*, 12785–12789. [CrossRef]
38. Sayen, S.; Carlier, A.; Tarpin, M.; Guillon, E. A Novel Copper(II) Mononuclear Complex with the Non-Steroidal Anti-Inflammatory Drug Diclofenac: Structural Characterization and Biological Activity. *J. Inorg. Biochem.* **2013**, *120*, 39–43. [CrossRef]
39. Lee, S.; Kapustin, E.A.; Yaghi, O.M. Coordinative Alignment of Molecules in Chiral Metal-Organic Frameworks. *Science* **2016**, *353*, 808–811. [CrossRef]
40. Paul, M.; Sarkar, K.; Deb, J.; Dastidar, P. Hand-Ground Nanoscale ZnII-Based Coordination Polymers Derived from NSAIDs: Cell Migration Inhibition of Human Breast Cancer Cells. *Chem.—A Eur. J.* **2017**, *23*, 5736–5747. [CrossRef]
41. Dilek, E.; Caglar, S.; Erdogan, K.; Caglar, B.; Sahin, O. Synthesis and Characterization of Four Novel Palladium(II) and Platinum(II) Complexes with 1-(2-Aminoethyl)Pyrrolidine, Diclofenac and Mefenamic Acid: In Vitro Effect of These Complexes on Human Serum Paraoxanase1 Activity. *J. Biochem. Mol. Toxicol.* **2018**, *32*, e22043. [CrossRef]
42. Braga, D.; Maini, L.; Grepioni, F. Mechanochemical Preparation of Co-Crystals. *Chem. Soc. Rev.* **2013**, *42*, 7638–7648. [CrossRef]
43. Karki, S.; Friščić, T.; Jones, W.; Motherwell, W.D.S. Screening for Pharmaceutical Cocrystal Hydrates via Neat and Liquid-Assisted Grinding. *Mol. Pharm.* **2007**, *4*, 347–354. [CrossRef]
44. Perumalla, S.R.; Pedireddi, V.R.; Sun, C.C. Protonation of Cytosine: Cytosinium vs. Hemicytosinium Duplexes. *Cryst. Growth Des.* **2013**, *13*, 429–432. [CrossRef]
45. Childs, S.L.; Stahly, G.P.; Park, A. The Salt-Cocrystal Continuum: The Influence of Crystal Structure on Ionization State. *Mol. Pharm.* **2007**, *4*, 323–338. [CrossRef]
46. Perlovich, G. Melting Points of One- and Two-Component Molecular Crystals as Effective Characteristics for Rational Design of Pharmaceutical Systems. *Acta Cryst. B Struct. Sci. Cryst. Eng. Mater.* **2020**, *76*, 696–706. [CrossRef]
47. Mukherjee, A.; Tothadi, S.; Chakraborty, S.; Ganguly, S.; Desiraju, G.R. Synthon Identification in Co-Crystals and Polymorphs with IR Spectroscopy. Primary Amides as a Case Study. *CrystEngComm* **2013**, *15*, 4640–4654. [CrossRef]
48. Heinz, A.; Strachan, C.J.; Gordon, K.C.; Rades, T. Analysis of Solid-State Transformations of Pharmaceutical Compounds Using Vibrational Spectroscopy. *J. Pharm. Pharmacol.* **2009**, *61*, 971–988. [CrossRef]
49. Glomme, A.; März, J.; Dressman, J.B. Comparison of a Miniaturized Shake-Flask Solubility Method with Automated Potentiometric Acid/Base Titrations and Calculated Solubilities. *J. Pharm. Sci.* **2005**, *94*, 1–16. [CrossRef]
50. Nugrahani, I.; Kumalasari, R.A.; Auli, W.N.; Horikawa, A.; Uekusa, H. Salt Cocrystal of Diclofenac Sodium-l-Proline: Structural, Pseudopolymorphism, and Pharmaceutics Performance Study. *Pharmaceutics* **2020**, *12*, 690. [CrossRef]
51. Putra, O.D.; Umeda, D.; Nugraha, Y.P.; Furuishi, T.; Nagase, H.; Fukuzawa, K.; Uekusa, H.; Yonemochi, E. Solubility Improvement of Epalrestat by Layered Structure Formation: Via Cocrystallization. *CrystEngComm* **2017**, *19*, 2614–2622. [CrossRef]
52. Putra, O.D.; Umeda, D.; Fujita, E.; Haraguchi, T.; Uchida, T.; Yonemochi, E.; Uekusa, H. Solubility Improvement of Benexate through Salt Formation Using Artificial Sweetener. *Pharmaceutics* **2018**, *10*, 64. [CrossRef]
53. Sanphui, P.; Rajput, L. Tuning Solubility and Stability of Hydrochloro-Thiazide Co-Crystals. *Acta Crystallogr. Sect. B Struct. Sci. Cryst. Eng. Mater.* **2014**, *70*, 81–90. [CrossRef]
54. McNamara, D.P.; Childs, S.L.; Giordano, J.; Iarriccio, A.; Cassidy, J.; Shet, M.S.; Mannion, R.; O'Donnell, E.; Park, A. Use of a Glutaric Acid Cocrystal to Improve Oral Bioavailability of a Low Solubility API. *Pharm. Res.* **2006**, *23*, 1888–1897. [CrossRef]
55. Childs, S.L.; Chyall, L.J.; Dunlap, J.T.; Smolenskaya, V.N.; Stahly, B.C.; Stahly, G.P. Crystal Engineering Approach to Forming Cocrystals of Amine Hydrochlorides with Organic Acids. Molecular Complexes of Fluoxetine Hydrochloride with Benzoic, Succinic, and Fumaric Acids. *J. Am. Chem. Soc.* **2004**, *126*, 13335–13342. [CrossRef]
56. Sathisaran, I.; Dalvi, S. Engineering Cocrystals of Poorly Water-Soluble Drugs to Enhance Dissolution in Aqueous Medium. *Pharmaceutics* **2018**, *10*, 108. [CrossRef]

MDPI
St. Alban-Anlage 66
4052 Basel
Switzerland
Tel. +41 61 683 77 34
Fax +41 61 302 89 18
www.mdpi.com

Crystals Editorial Office
E-mail: crystals@mdpi.com
www.mdpi.com/journal/crystals

www.ingramcontent.com/pod-product-compliance
Lightning Source LLC
LaVergne TN
LVHW070646100526
838202LV00013B/892